Pastoral Politics

Association of Ancient Historians

The purpose of the monograph series is to survey the state of the current scholarship in various areas of ancient history.

Other publications by the Association

PASTORAL POLITICS

ANIMALS, AGRICULTURE AND SOCIETY IN ANCIENT GREECE

Publications of the
Association of
Ancient Historians 9

Timothy Howe
St. Olaf College

Regina Books
Claremont, California

Book design: Mark Morrall Dodge
Cover design: Mary Stoddard

ISBN: 1-930053-54-1 and 978-1-930053-54-0

Library of Congress Control Number: 2008920959

Co-published by arrangement with the
Association of Ancient Historians

𝕽𝖊𝖌𝖎𝖓𝖆 𝕭𝖔𝖔𝖐𝖘
Post Office Box 280
Claremont, California 91711
Tel: (909) 624-8466 / Fax (909) 626-1345
Website: www.reginabooks.com

Manufactured in the United States of America

to my wife,
Mary,
sine qua non

CONTENTS

PREFACE

When Carol Thomas first suggested to me at the 2005 Annual Meeting of the Association of Ancient Historians in Columbia, MO, that I consider contributing a volume to this series about the interdependencies between land use, animals, agriculture and politics in ancient Greece, I was honored, excited, and challenged all at once. In keeping with the goals of the series, the book would need to provide an overview of the interactions between animals, land and agriculture to ancient historians who had little or no knowledge of the subject—a daunting task in and of itself. But I also felt the book would need to justify why ancient historians should care about animals and agriculture—and here I was especially challenged. I had written about animal husbandry before, and this work reflects some of those earlier interests and thinking, yet animal husbandry itself seemed too narrow, especially for something that should be accessible to non-specialists, hopefully spark wider debate and even provide a pedagogical resource for a subject traditionally isolated from mainstream historical debates.

It seemed to me that the way forward was through a discussion of land use, especially politicized land *non-use*, and decided to start with three simple questions: (1) why did wealthy (and even some non-wealthy) people in a dry, mountainous region like Greece prioritize the production of animals to such a degree that they removed some of the best land from cereal or other food cultivation; (2) how did these people justify taking much needed land away from subsistence food production in order to raise non-food animals such as horses; and (3) how did these animal production choices affect those individuals directly and *not directly* involved in animal production? What follows is my attempt to answer these questions and show that Greek choices about animal production and animal consumption

affected ancient peoples at all levels of society in a multitude
of ways.

This monograph owes a great deal to a great many. My
parents, Richard and Carole, gave me years of practical
guidance in the keeping of horses, sheep, goats and cattle in
the foothills of Northern California's Sierra Nevada
mountains, and without their help I would never have been
able to make sense of ancient herding practice. My mentor,
Mark Munn, gave me the tools, guidance and, above all, the
encouragement to tackle the nuances of Greek rural history.
My wife, Mary, fought me tooth and nail over every draft
and forced me to justify every argument; if the writing is
clear, even to the non-specialist, then much of the credit goes
to her. My colleagues in the History and Classics
departments at St. Olaf, who were always supportive,
especially when I was battling a tough Greek inscription, and
who kindly did not look too puzzled when I told them I was
writing about animals. The 18 St. Olaf students in my
Spring, 2007 senior research seminar on Archaic Greece, who
helped me work through much of the theoretical literature
about early Greece, challenged many of my heartfelt
assumptions and in so doing forced me to justify and refine
my views on Archaic and Classical politics.

But perhaps the greatest thanks goes to Carol Thomas,
Jennifer Roberts and Lee Brice, for reading an earlier draft,
helping to sharpen my argument and making the work more
accessible. At every step of the way these members of the
AAH publication committee were supportive and
encouraging. Of course, any errors of fact or judgment are
mine alone.

I

UNDERSTANDING PASTORAL POLITICS

In 370 BCE, Jason, tyrant of Pherai and warlord of all Thessaly, ordered his subjects to provide one thousand cattle, along with over ten thousand sheep, goats and swine at his capital of Pherai, so that he could make an offering to Delphian Apollo during the upcoming pan-Hellenic Pythian Games. In an effort to incite competition among his Thessalians, and thus gain animals of the highest quality, he even offered a golden crown to whoever provided the finest bull. Jason's goal: a huge sacrificial procession, complete with herdsmen, guards and camp followers marching from Pherai, down through the communities of central Greece to Apollo's sanctuary at Delphi. To make the spectacle even more impressive, and all the more secure, Jason commanded many of his Thessalian subjects to take the field with him and the animals, march to Delphi, and attend the games. Why? What could induce anyone, let alone a Thessalian warlord, to make such a large animal dedication at a shrine so far from home, in the presence of both his own countrymen and an international audience of other Greeks? What did he hope to communicate by assembling and marching these animals across Central Greece and then having them slaughtered in a public spectacle? And why did Jason need so many animals? For that matter, how could one man have access to over 10,000 *surplus* beasts, especially 1,000 full-grown, expensive-to-produce beasts like cattle, and still maintain the integrity of his herds? The answers to these questions are complex and encompass a network of human choices, values, and behaviors that I call "pastoral politics."

Ancient Greeks from the elite to the wretchedly poor inherently understood the answers to the questions asked above. As cultural insiders they understood pastoral politics. Greeks from Thessaly to Sicily could decode the symbolism conveyed by animal wealth. And that's part of the problem for the modern researcher, who is anything but an insider and consequently has to rely on native informants like the classical authors. But the ancient literary sources are not the most useful informants because they do not, on the whole, explain what should be obvious to their ancient readers. Consider, for example, Jason's contemporary, the Athenian historian Xenophon, who recorded the tyrant of Pherai's actions in the sixth book of his *Hellenica*. Xenophon tells his readers explicitly what Jason was doing and why he did it in terms they would understand, without any unnecessary explanation: Jason organized his animal parade and sacrificial spectacle because he wanted to become director of both the feast in honor of Delphian Apollo and of the Pythian Games which followed (*Hell*. 6.4.29). A terse and compressed explanation that leaves the modern reader asking: how and why would a sacrificial dedication do that? How could animals gain Jason an important religious-political office like Director of the Pythian Games? Of course, Xenophon did not need to explain the mechanics of the situation, the how and why, to his classical readers because like himself his audience were native speakers of pastoral politics and would understand intimately not only how such an expensive display and dedication of animals was possible for someone like Jason but also how it could result in appointment to a position of public authority like Director of the Games. The key to unlocking pastoral politics for the *modern* reader, and a major burden of this study, lies in decoding the ancient Greek values behind large-scale animal production, as well as the social and political bonds of reciprocity and obligation that animal displays create. Unlocking pastoral politics can unlock a whole network of human choices and relationships and thereby greatly expand our understanding of ancient Greek culture.

Let's return to the example of Jason and his Delphic dedication. The tyrant of Pherai was able to assemble so many surplus animals because Thessaly, unlike many regions in Greece (e.g. Attika), has abundant, naturally irrigated pastures as well as extensive arable land for cereal production. Consequently, in Thessaly grazing does not compete with farming for land and water. But even if grazing did compete with agriculture, grazing would win out because the Thessalians valued animal wealth over all other types and therefore gave surplus animal production greater priority in terms of land use and labor. Moreover, Thessaly's unique social institutions, which allowed an aristocratic elite to rule over a dependent, serf-like workforce, supplied the thousands of watchmen, shepherds, cowherds, and swineherds necessary for the Thessalian lords to engage in such large-scale, non-subsistence animal production. As warlord of Thessaly, Jason was the top "land lord" and because of the social framework of the region, could command or inspire gifts of animal wealth from his underlings. In fact, he did this in a very clever way—he inspired the Thessalian elites to compete for public recognition and status. The lords of Thessaly competed among themselves to provide Jason with the best bull for his animal dedication. The winner received a golden crown from Jason and was thereby publicly acknowledged as the best stockman in all of Thessaly. Given how Thessalians prioritized animal production above other types of wealth, supplying the best animal for a tremendous spectacle like Jason's would be a high honor and a serious accomplishment. For these social and political reasons Jason had access to so many surplus animals.

But why did he want to march them out of Thessaly and sacrifice them to a god in front of both Thessalians and foreigners? Why not just sacrifice them in Thessaly? In having his demands for over 10,000 animals met by his subjects, Jason had already proved he was preeminent at home. His audience, and therefore his goal, was far larger than just Thessaly. By expending his own resources on behalf of such a regionally important god, in the form of sacrificial victims, Jason would send a clear message about his *regional*, even international preeminence. Jason chose Delphi both because of its Mediterranean-wide reputation as an oracular site and because of the well-attended upcoming Pythian Games which it hosted. By providing *all* of the required sacrificial victims for the Pythian festival, Jason placed the god *and* the Delphic priests in his debt, a debt that Delphi could easily repay by granting him the Pythian directorship he sought. And this directorship was no empty honor, for the Director of the Pythian Games held real political and social power because he presided over the various competitive events and proclaimed the winners of each. In a sense, the Director (in this case Jason) was be Apollo's foremost representative, bestowing *kudos* and honor to the victors on the god's behalf. By the display of divine authority made possible from the Directorship of the Games, and through the sacrifice itself, Jason gained direct influence over the assembled audience. He was the leader, he was the "master of ceremonies," and he was *officially* in control of the audience and participants of the Games.

The sacrifice itself only intensified this impression of official power. In addition to the god and his priests, the animal dedication also placed the spectators under an obligation. As witnesses to the sacrifice and participants in the ritual feast of meat that followed, the assembled audience would partake personally in Jason's gift and consequently feel a debt of gratitude to their benefactor for the free meal they had received. This obligation operated only on a subliminal level but was powerful just the same. Jason gave while the god and the audience received. In addition, and far more importantly, on a practical level beyond the symbolism of gift and obligation, the thousands who attended the festival and the Games would have been awed by the spectacle of so many animals belonging to one man being dispatched. From that moment, the audience would be convinced that Jason was a man of tremendous resources, power, and wealth. That big impression, that destruction of surplus wealth on such a grand scale, was the main purpose of this conspicuous consumption of animals in an internationally visible location like Delphi. Indeed, the fact that Jason intended to parade these thousands of beasts through the lanes of central Greece on his way from Pherai to Delphi, rather than simply purchase them locally, suggests that impression underpinned the entire spectacle, from start to finish.[1] The animal procession to the Pythian Games could best demonstrate to all who witnessed it (or even just heard about it secondhand) that Jason truly was pre-eminent and thus worthy to be *hegemon* of all Greece, or at least that he was far too powerful to oppose. And it probably would have worked. Jason probably would have become *hegemon* of the central Greeks because of his animal display, had he not been killed before the animals ever left Thessaly, stabbed in the back by a paid assassin as he was reviewing and inspecting his Pheraian cavalry.

Apart from his untimely death, Jason's attempt at pastoral politics was all too common among wealthy ancient Greeks and unfortunately has been all too commonly overlooked by modern researchers. This is not to say that scholars of the ancient world have not long recognized the necessity of animals in public events such as feasts, sacrifices and games. For over a century, scholars of Greek religion and public events have documented the use of animals (often donated by wealthy citizens like Jason) as sacrificial victims,[2] or as the main course at public feasts (also sponsored by the wealthy),[3] or even as the main attraction at elite public competitions such as the horse and

---

[1] Diodoros (15.60.1-2) implies that impressing the neighbors was an issue.

[2] E.g. W. Burkert, *Homo Necans* (Berekeley and Los Angeles: University of California Press, 1983), and *Greek Religion* (Cambridge, MA: Harvard University Press, 1985); and V. Rosivach, *The System of Public Sacrifice in Fourth-Century Athens* (American Classical Studies 34) (Atlanta: Scholars Press, 1994).

[3] E.g. M. Detienne, Marcel and J.-P. Vernant, edd., *The Cuisine of Sacrifice among the Greeks* (Chicago: University of Chicago Press,1989); G. Berthiame, *Les rôles du mágeiros. Étude sur la boucherie, la cuisine et le sacrifice dans la Grèce ancienne* (*Mnemosyne* Suppl. 70) (Leiden: Brill, 1982); and V. Rosivach, *Public Sacrifice*.

chariot races of the pan-Hellenic Pythian Games.[4] In addition, works about ancient animal production have catalogued extensively what animals were raised in what regions by means of what production strategies.[5] Nonetheless, these same scholars have shown a general reluctance to move beyond the boundaries of the subject at hand and explore the wider context of animal production and animal display. Little attention has been given to pastoral politics, to putting together in an interpretative framework both why and how the ancient Greeks produced animals on a large scale, especially expensive-to-maintain animals such as cattle and horses. Even less attention has been centered on understanding the interdependencies between elite need for animal production and their decisions about labor, land use, land acquisition, and land protection in the form of laws and wars.[6] Therefore, a major burden of this book will be to unite the subfields of animal management and animal display and thereby bring animal husbandry into the mainstream historical discussion and sort out not only how and why wealthy and powerful men such as Jason of Pherai produced animals and used them in costly displays, but also to what extent such non-subsistence animal production in a land dominated by seasonal drought and mountainous geography affected and shaped Greek communities over time. To put it another way, this study will illustrate to what extent the demands of pastoral politics shaped ancient

[4] E.g. R. A. Harris, *Greek Athletes and Athletics* (Bloomington, IN: Indiana University Press), *Sport in Greece and Rome* (Ithaca: Cornell University Press, 1972); D. C. Young, *The Olympic Myth of Greek Amateur Athletics* (Chicago: Ares Press, 1984); M. Golden, *Sport and Society in Ancient Greece* (Cambridge: Cambridge University Press, 1998); and D. Kyle, *Sport and spectacle in the ancient world* (Malden, MA: Blackwell, 2007).

[5] E.g. K. Winkelstern, *Die Schweinezucht im klassischen Altertum* (Diss., Giessen, 1933); K. Zeissig, *Die Rinderzucht im alten Griechenland* (Diss., Giessen, 1934); O. Brendel, *Die Schafzucht im alten Griechenland* (Diss., Giessen, 1934).

[6] C. Chandezon, *L'élevage en Grèce (fin Ve-fin Ier s.a.C.). L'apport des sources épigraphiques* (Bordeaux: Ausonius, 2003), is a notable exception, though he is concerned only with the late Classical and Hellenistic periods and limits his analysis to national or communal trends rather than individual behavior. See discussion below.

Greek life in general, for all Greeks both urban and rural, elite and non-elite, during the Archaic and Classical and early Hellenistic periods.

While this focus on how human choices about raising animals affected and were affected by human institutions is new, an interest in the agricultural and pastoral activities of humans in the ancient world is not. Attention to the interdependencies between agriculture and society really began with the works of Louis Gernet and Auguste Jardé, and centered around the role of arable agriculture in the subsistence strategies of the ancient Greeks, especially the Athenians.[7] Gernet, for example, was the first to probe the interdependencies between the settled population and the land by examining the demands of the urban food supply. Even though he lacked the comparative data of population demographics now available, Gernet was able to make some sophisticated and far-reaching observations about the inability of Attika to feed the large urban population of fifth- and fourth-century Athens and the resultant heavy dependence of the Athenians on a foreign grain supply, that have shaped almost a century of scholarship.[8] In a similar fashion, Jardé laid the foundation for our present understanding of the mechanics of ancient cereal production, demonstrating that grain was the cornerstone of ancient subsistence food production, and a primary concern to all ancient Greek agricultural strategies.

From this agricultural framework, M. I. Finley built his model of the ancient Mediterranean economy, confirming that ancient economic activity centered primarily around the needs of subsistence food production and surplus market production.[9] Finley constructed a household, agricultural model driven by personal needs because the

[7] L. Gernet, *"L'Approvisionnement d'Athènes en blé au Ve et au IVe siècle,"* Université de Paris, Bibl. Fac. Lett. 25 (Paris: Mélanges d'histoire ancienne, 1909), and A. Jardé, *Les Céreales dans l'antiquité greque* (Paris, 1925).

[8] R. Sallares, *The Ecology of the Ancient Greek World* (Ithaca, New York: Cornell University Press, 1991), for a review of the relevant literature.

[9] M. I. Finley, *The Ancient Economy* (Berkeley and Los Angeles: University of California Press, 1985).

lack of ancient literary treatises on economic theory suggested to him that the Greeks and Romans had no real conception of the "economy" as such, other than at the household level, and certainly nothing as nuanced or complex as some modern economic systems like capitalism or Marxism. According to Finley, instead of profit-margins and modes of production, the ancient Greek or Roman thought first about his stomach, about bread and meat for today and tomorrow. Only after he had satisfied his subsistence food needs could the ancient producer then begin to think about surplus, whether agricultural, pastoral, commercial, or industrial. Once subsistence goals were met, the ancient Greek or Roman could then exchange surplus agricultural products for luxuries that would make life more amenable in general. Very rarely would a wealthy man plan to increase his yield by re-investing profits, for unlike the modern man of wealth, his ancient counterpart had no set ideology which urged that capital be continually re-invested in order to gain ever higher returns. Comfort and status were the goal, and the unlimited pursuit of wealth production, simply for the sake of wealth production, had little honor or respect in the ancient world and was regularly condemned by ancient social theorists.[10] Wealth served a social purpose and required a social outlet such as the gift of a chorus, a warship, a temple, or an animal sacrifice. In this way, wealth created goodwill, and goodwill created status, social prominence, and above all security.[11]

[10] Aristotle, in both the *Politics* and the *Nikomachian Ethics* devotes much space to the "proper" uses of wealth. For him, unlimited wealth acquisition is especially vulgar. The idea that wealth usually had a social dimension, a purpose, among the ancients is an idea to which we shall return (See discussions in chapters 2 and 5, below).

[11] By touching on these social uses of wealth Finley is building of the foundations laid by E. Durkheim, *De la division du travail social* (Paris: F. Alcan, 1911); M. Mauss, "Essai sur le don. Forme et raison de l'échange dans les sociétés archaïques" *l'Année Sociologique*, seconde série, 1923-1924 [W. D. Halls, trans., *The gift: the form and reason for exchange in archaic societies* (New York: W.W. Norton, 2000]; and K. Polyani, *The great transformation* (New York: Farrar & Rinehart, 1944). See P. Veyne, *Bread and circuses: historical sociology and political pluralism* (London: Penguin, 1990), for a recent treatment of these dynamics.

Although Finley tended to paint a very general picture of the ancient economy, often treating the Greek and Roman worlds as a single analytical unit, his view of ancient Mediterranean life as regulated and shaped by considerations of subsistence food production and surplus market production still dominates most models of ancient socio-economic behavior.[12] And from this theoretical foundation a group of scholars began to explore the nuances of rural agricultural production and urban demand. One of first was Peter Garnsey, who challenged the prevailing view, originally advanced by Gernet, that during the fifth and fourth centuries BCE the Greek countryside was unable to feed its urban areas.[13] By examining in some detail various Greek and Roman responses to chronic food shortage, as well as ancient systems of food production, Garnsey was able to demonstrate that, while food shortage was common, true famine was relatively rare, the outcome of abnormal conditions. On the whole the rural landscape was able to produce sufficient food to support both rural and urban inhabitants.[14] In Greece, for example, only fifth-century Athens, with its urban population swollen and cut off from the produce of the countryside by the exigencies of the Peloponnesian War, truly needed to import food, and then only for the duration of that particular war.[15] The Athenians, in the main, were not tied to the international

[12] E.g. the essays collected in Walter Scheidel and Sitta von Reden, edd., *The Ancient Economy* (London: Rouledge, 2002) and Paul Cartledge, ed., *Money, labour, and land: approaches to the economies of ancient Greece* (London: Routledge, 2002).

[13] P. Garnsey, *Famine and the Food Supply in the Graeco-Roman World* (Cambridge: Cambridge University Press, 1988).

[14] See the essays collected in P. Halstead and J. O'Shea, edd., *Bad year economics : cultural responses to risk and uncertainty*, (Cambridge: Cambridge University Press, 1989), for similar conclusions.

[15] T. W. Gallant, *Risk and Survival in Ancient Greece. Reconstructing the Rural Domestic Economy* (Stanford: Stanford University Press, 1991), whose work focused more closely on the rural domestic production of Ancient Greece, came to similar conclusions. Garnsey followed up his earlier work with a book devoted to food and its role in ancient Mediterranean society. P. Garnsey, *Food and Society in Classical Antiquity* (Cambridge: Cambridge University Press, 1999).

grain market and its vagaries because of local food shortage, as Gernet and others had supposed. Instead, Athens had chosen to become dependent on the import grain market because of complex social and political decisions by the ruling elite. For the first time, a clear argument was advanced that elite politics and personal choice, rather than the "natural laws" of supply and demand, or Finley's subsistence food production and surplus market production, governed agricultural practice. According to Garnsey, the Athenians of the later fifth century BCE *chose* not to cultivate their land and instead import grain in order to engage in a particular political strategy, relocation behind the Long Walls. During the stresses of the Peloponnesian War, the need for a less complex local solution to subsistence food production needs did not take priority over political concerns, and the Athenians were content to seek grain beyond their sphere of control, even though this meant that outside forces could manipulate Athens' food supply.[16] Garnsey's conclusion introduced the social variable of political choice and thereby opened the way for a more nuanced consideration of ancient Greek economic and agricultural behavior.

Like Garnsey, Robin Osborne attacked another long-held, pervasive agricultural model: the artificial division between city and countryside. In his work on Attika, Osborne examined the territory of the Athenians in terms of the connections between the ancient city, its needs, and the rural hinterland.[17] What emerged was a view of city and countryside, of Athens and Attika, joined by overlapping bonds of mutual cooperation, not divided by competition or separated into defined spheres of influence. A few years later, Osborne applied this method on a larger

---

[16] Victor Hanson, *The Other Greeks* (Berkeley and Los Angeles: University of California Press, 1999), disagrees and argues that local control over production and self-sufficiency are the highest concern for the ancient Greek farmer. Only if faced with no other choice will a Greek surrender his ownership over the food supply.

[17] R. Osborne, *Demos: The Discovery of Classical Attika* (Cambridge: Cambridge University Press, 1985).

scale to other areas of Greece.[18] The result was a general
discussion of the Greek landscape beyond Attika, focused
on the "exchange" of products and services. Here, the
connections between the rural and urban worlds were
accentuated and explained in terms of complementary
obligations, instead of rivalries or competition for
resources.[19] By exploiting both epigraphic and literary
sources, Osborne demonstrated the interactive role of the
urban population in the world beyond the city walls. Both
urban and rural Greeks participated regularly in
community-wide festivals and sacrifices in the agricultural
life of the family farm and orchard, and even in warfare
and border defense in the uninhabited hinterlands. By
stressing the "agricultural" calendar, with its rural festivals
and processions, and the everyday role of livestock in blood
sacrifice, as intermediaries between men and gods, Osborne
showed how close the agricultural world was for the urban
population and how far both urban and rural choices might
affect both worlds. Indeed, the powerful elite, by dictating
festivals, sacrifices, and wars determined how the
countryside might be used. Yet the countryside was never
just an agricultural resource, subject to the demands and
whims of an urban population. The rural landscape was as
essential and integrated a part of citizen life and citizen
decision making as the agora because the ruling elites, as
rural landholders, were anchored to the countryside and its
realities as much as they were to the urban center and their
townhouses.[20] According to Osborne elite political choices

[18] R. Osborne, *The Classical Landscape with Figures* (London: George Philip, 1987).

[19] The Copenhagen Polis Centre, under the direction of Mogens H. Hansen, has been exploring a similar line of research. See, for example, M. H. Hansen, ed., *The Polis as an Urban Centre and as a Political Community, Acts of the Copenhagen Polis Centre 4* (Copenhagen: KDVS, 1999); M. H. Hansen and T. H. Nielsen, edd., *An inventory of archaic and classical poleis* (Oxford: Oxford University Press, 2004); and M. H. Hansen, *The shotgun method : the demography of the ancient Greek city-state culture* (Columbia: University of Missouri Press, 2006).

[20] See R. M. Rosen and I. Sluiter, edd., *City, Countryside, and the Spatial Organization of Value in Classical Antiquity. Mnemosyne Supplements* 279 (Leiden: Brill, 2006), for the most recent discussions of rural and urban interdependencies.

shaped agricultural production in the same way that the
realities of agricultural production, in turn shaped political
choices.

Perhaps taking a cue from scholars such as Osborne,
and certainly dissatisfied with the isolation of Greek
agricultural scholarship from recent advances in fields such
as population demographics, social anthropology and
agricultural history, Robert Sallares attempted to
contextualize further the world of the Ancient Greek
community by exploring the fundamental interdependence
of plants, animals, humans, climate, and geology—in short,
what Sallares chose to call the ancient "ecology."[21]
Through the use of comparative ethnographies from East
Africa and New Guinea, Sallares explored the impact of
humans on the ancient Greek world. As a result, he opened
up a vast technical literature previously opaque to scholars
of the ancient world, and made it difficult for anyone in
future to suggest that the ancient city was only loosely
connected to the rural countryside. Although Sallares'
arguments depend heavily on complex models of
population demography, and many of his conclusions have
been disputed by more traditional ancient historians,[22] his
highly technical discussion of the evolution of cereal crops
in the first millennium BCE (the last third of the work) is a
significant contribution to our understanding of ancient
subsistence food production. In this section, Sallares broke
down subsistence into its components and assessed the
roles and evolution of each plant and animal as a
domesticated species. His conclusion that ancient
subsistence food production strategies centered primarily
on grains should come as no real surprise to readers of
Jardé, though the scientific weight Sallares brings to this
conclusion is welcome and should settle the question of
subsistence for some time to come. The corollary to this
conclusion, however, is that animals contributed little if at
anything to ancient subsistence. Sallares' analysis
confirmed that animal production for the ancient Greeks

[21] R. Sallares, *The Ecology of the Ancient Greek World*.
[22] Most notably, Victor Hanson in *The Other Greeks*.

was primarily a surplus activity and suggests that we look beyond subsistence food production for explanations for why the Greeks raised animals.

At the same time that Gernet and Jardé were working out the agricultural systems of Classical Athens, scholars at Giessen in Germany were cataloguing, species by species, the types of animals the ancient Greeks produced, and the historical geographer E. C. Semple was synthesizing a general picture of how such animals were raised.[23] While reading the ancient testimonia, Semple noticed many parallels between ancient descriptions and the animal production methods of contemporary Mediterranean stockmen. These many similarities across time and cultural boundaries led Semple to conclude that some factor or factors, common to all must have a dominant influence on the character of Mediterranean animal production.[24] Semple argued that the Greek landscape of high mountains and coastal plains, and climate of hot dry summers and warm wet winters permitted only a limited type of animal production—a seasonal shifting of sheep and goats between winter lowland and summer upland pastures, known to specialists as transhumance. Herds of large animals such as cattle and horses could flourish only in the few areas of

---

[23] The 1920s and 1930s produced a number of specialized treatises on ancient animals. K. Winkelstern, *Die Schweinezucht im klassischen Altertum*; K. Zeissig, *Die Rinderzucht im alten Griechenland*; O. Brendel, *Die Schafzucht im alten Griechenland*. E. C. Semple, "The Influence of Geographic Conditions upon Ancient Mediterranean Stock-Raising," *Annals of the Association of American Geographers* 12 (1922): 3-38, and *The geography of the Mediterranean region and its relation to ancient history* (New York: H. Holt and Company, 1932).

[24] Practices of "traditional" societies are used much more cautiously by today's archaeologists and ethnologists. See for discussion Claudia Chang, "Ethnoarchaeological survey of pastoral transhumance sites in the Grevena region, Greece" *Journal of Field Archaeology* 20 (1993): 249-264; and "Pastoral transhumance in the Southern Balkans as a social ideology: ethnoarchaeological research in northern Greece," *American Anthropolgist* 95 (1993): 687-703. P. Halstead, in "Present to Past in the Pindhos: Diversification and Specialization in Mountain Economies," *Rivista di Studi Liguri* 56 (1990): 61-80, and "Pastoralism or household herding? Problems of scale and specialization in early Greek animal husbandry," *World Archaeology* 28 (1996): 20-42, explains the dangers inherent in using enthographies in this way.

perennial wetland, such as the Kopais and Thessalian basins, the upland lakes of Arkadia, or the fertile river valleys of Lakonia, Messenia and Elis, which supplied plentiful, year-round grazing.

Because of this environmentally determined methodology, where animal management strategies resulted largely from topographical and climatic imperatives, some scholars have criticized the value of Semple's thesis.[25] Yet the thrust of her argument, that environment and topography affect systems of animal husbandry is valuable and should not be dismissed out of hand: herds of the larger domestic animals (more than 2 cattle) need areas of irrigated pasturage, or regular supplemental fodder such as agricultural waste or hay, while large herds of smaller stock like sheep and goats need supplemental feeding or regular movement among otherwise useless "scrublands."[26] Semple's conclusions about the limiting factors of climate and topography upon herd size, herd type (small or large livestock), and animal production strategies are her real contribution and did much to advance understanding of the realities of animal production in a dry region like Greece.

After the works of Semple and the "Giessen School," ancient Greek animal husbandry attracted little scholarly attention until the early 1970s, when Stella Georgoudi returned to Semple's environmental model and applied it to a new body of evidence—the epigraphic corpus.[27] Using the epigraphic studies of Robert alongside the literary evidence, Georgoudi sought to prove more conclusively that ancient Greek stockmen systematically moved their animals from season to season in an effort to exploit the cool, abundant summer pastures of the mountains and the well-watered

---

[25] See discussion below.

[26] O. Rackham, "Observations on the Historical Ecology of Boeotia," *ABSA* 78 (1983): 291-351.

[27] S. Georgoudi, "Quelque problèmes de la transhumance dans la Grèce ancienne," *REG* 87 (1974): 155-85.

winter pastures of the sheltered lowland plains.[28] Georgoudi argued that numerous ancient references to transhumance suggest that this mobile form of large-scale animal production was *the* prevailing management strategy for sheep and goats among the ancient Greeks.[29] Her thorough analysis of the available literary and epigraphic data firmly established that transhumance was known and practiced from an early date (at least by the late 800s BCE) in *some* areas of ancient Greece, by *some* Greeks.

A challenge to the environmental model came late in the 1980s when Paul Halstead brought new evidence and a new perspective to bear.[30] Halstead, like Osborne, Garnsey and Sallares was part of a movement of scholars seeking more complex and nuanced social and cultural explanations for ancient behaviors. Through an analysis of the archaeological and literary evidence, Halstead argued that the ancient Greeks did not have the market demand or access to land that could support purely pastoral strategies such as long-distance transhumance. Consequently, no one could specialize in raising *just* animals. Producers had the mitigate risk by raising plants and animals in tandem and by keeping their stock near their farms in order to achieve the greatest and most efficient use of available resources. Keeping animals on or near the farm base allowed farmers to convert otherwise useless waste products, such as stubble, cover crops and olive pressings into usable, saleable goods such as wool, hides and meat.[31] To turn herds out into distant summer and winter pastures was not an economically desirable strategy because it wasted not only a large amount of resources in transit and labor, but

---

[28] L. Robert, *Études épigraphiques et philologiques* (Paris, 1938); idem, "Epitaphe d'un berger á Thasos," *Hellenika* 7 (1949): 152-160; idem, "Les chèvres d'Herakleia," *Hellenika* 7 (1949): 161-170; and idem, "Monnaies d'Olympos," *Hellenika* 10 (1955): 185-186.

[29] Cattle and horses seem to have little place in Georgoudi's model of ubiquitous transhumance.

[30] P. Halstead, "Traditional and Ancient Rural Economy in Mediterranean Europe: Plus Ça Change?" *JHS* 107 (1987): 77-87.

[31] It is clear from his discussion, however, that Halstead is primarily interested in animal production on the small, household, level.

also resulted in the loss of half a year's worth of farm fertilizer in the form of manure.[32] In the end, the sort of specialized transhumance proposed by Georgoudi and Semple, which involved a large degree of seasonal movement and disconnection from agriculture, was by no means a normal or "natural" response to the environment. Indeed, large-scale seasonal transhumance could exist only if stockmen had uninhibited access to pasture, reliable control of labor for herdsmen and guards, and above all steady, dependable markets for pastoral products like cheese, wool, hides, and meat that continually rewarded investments in pastoral infrastructure—none of which was present in ancient Greece at any time in antiquity.[33]

In addition to refuting the environmental model's conclusions, Halstead also critiqued its observations about an unchanged landscape:

> The present summer pastures in the mountains are, to a large extent, not a 'natural' feature of the Mediterranean landscape. Although tree growth may be prevented locally in the mountains by steepness of slope, absence of soil, waterlogging and so on, no Mediterranean mountain is high enough (for its southerly latitude) for extensive alpine meadows to be the inevitable product of harsh winter conditions...*Most mountain pasture seems to be the product of human interference*—either directly through fire and axe or indirectly through grazing livestock...Therefore, perhaps well into early historical times, mountain pasture may have been very limited in extent.[34] [emphasis added]

---

[32] Manure was a very desirable resource. See S. Hodkinson, "Animal Husbandry in the Greek Polis," in C. R. Whittaker, ed., *Pastoral Economies of Ancient Greece and Rome* (Cambridge Philological Society, Suppl. Vol. 41) (Cambridge: Cambridge University Press, 1988), 35-74.

[33] John Cherry, "Pastoralism and the Role of Animals in the Pre- and Protohistoric Economies of the Aegean," in C. R. Whittaker, ed., *Pastoral Economies of Ancient Greece and Rome* (Cambridge Philological Society, Suppl. Vol. 41) (Cambridge: Cambridge University Press, 1988), 196-209, reiterates these preconditions in his critique of Dark Age Greek nomadism.

[34] P. Halstead, "Traditional and Ancient Rural Economy," 79. For similar conclusions see, P. Garnsey, "Mountain Economies in Southern Europe," in C. R. Whittaker, ed., *Pastoral Economies of Ancient Greece and Rome*

And yet Halstead's own observations about a changed landscape are equally flawed, as they suffer from several unsupported presuppositions: (1) a classical landscape much more forested than now, with upland grazing created solely by human action; (2) plants and animals as passive recipients of whatever humans choose to inflict on them; and (3) an environment irrevocably changed by human activity. Humans certainly changed the Greek landscape over time by cutting back forests, clearing land for agriculture, building walls, terraces and roads, diverting streams and springs for irrigation, but a significant amount of upland pasture seems to have existed naturally by the time of our earliest written sources. Consider the Parnassos massif as an example.[35] As early as the Homeric epics, for example, large herds roamed the upland pastures of Parnassos: "In Phoibos Apollo's house on Pytho's [Parnassos'] sheer cliffs, many cattle and fat sheep can be had for the raiding" (Iliad 9.404-5). And just east of the Delphic shrine is the carefully demarcated territorial boundary of the Ambryssians and Phlygonians that snakes its way far up the slopes of the mountain.[36] The fact that these two communities bothered to mark a boundary in the upland areas, unprofitable for grain and olive production, given their short growing season, suggests both an interest in grazing and the presence of useable pasture by the 4th century BCE. Indeed, a passage from the Oxyrhynchos

---

(Cambridge Philological Society, Suppl. Vol. 41) (Cambridge: Cambridge University Press, 1988), 196-209; J. Lewthwaite, "Plains tales from the hills: transhumance in Mediterranean archaeology," in A. Sheridan and G. Bailey, edd., *Economic Archaeology: Towards an Integration of Ecological and Social Approaches* (BAR S96) (Oxford Clarendon Press, 1981), 57-66. But as Rackham has shown, Mediterranean trees are much more resilient than commonly thought. Many species cannot be wholly removed by woodcutting or browsing by animals. See O. Rackham, "Land use and the Native Vegetation of Greece," in M. Bell and S. Limbrey, edd., *Archaeological Aspects of a Woodland Ecology* (Oxford: BAR International Series, 1982),177-98; "Observations on the historical ecology of Boeotia," *ABSA* 78 (1983): 291-251; and O. Rackham and J. Moody, *The Making of the Cretan Landscape* (Manchester: Manchester University Press, 1996).

[35] See discussion of upland forest limitations below.

[36] *Fouilles de Delphes* iii. 2. 140-7.

Historian (*Hell. Oxy.* 18.3-4), makes such a reading of the Ambryssian-Phlygonian border all the more attractive, as a grazing dispute between the Phokians and Lokrians somewhere on the Parnassos massif (probably the side opposing Delphi), expanded into pan-hellenic war (the Korinthian War of 395 BCE). The heights of Parnassos show an early and intensive history of grazing.

Halstead's presupposition of a more-forested, greener ancient Mediterranean made bare by human activity has been around for so long that it has taken on the mantle of established fact. First put into words by Plato in antiquity (*Kritias*) and eloquently described by Sonnini in the modern period, the model of the "destroyed environment" is still very much alive, as Clive Ponting's bestselling, *A Green History of the World: the environment and the collapse of great civilizations* demonstrates.[37] According to Ponting and others, magnificent forests covered the Mediterranean countryside until men came and cut down the forests to clear land for farming and grazing. Then, once clear-cut, the trees failed to regenerate and both the land and the climate degraded into semi-desert. As a final insult, goats and sheep rapaciously grazed the resultant scrub and degenerate pasture, ensuring and accelerating the process of desertification.[38] While adherents of this theory accept that the process began in antiquity, they argue that the change only became critical in the modern period—the last 100 years or so. Consequently, centuries of increasingly destructive, human-initiated environmental degradation produced the landscape of "scrub" lowlands,

---

[37] C. S. Sonnini, *Voyage en Grèce et en Turquie fait par ordre de Louis XVI* (Paris, 1801); C. Ponting, *A Green History of the World: the environment and the collapse of great civilizations* (London: Sinclair-Stevenson, 1991). See also J. D. Hughes, *Pan's travail: environmental problems of the ancient Greeks and Romans* (Baltimore : Johns Hopkins University Press, 1994).

[38] While humans are largely responsible, domestic animals are their unwitting tools in the desertification of the Mediterranean. For this and other reasons, animal husbandry has received little positive attention in the mainstream literature over the years. See, for example, J. R. McNeil, *The Mountains of the Mediterranean World: an environmental history* (Cambridge: Cambridge University Press, 1992).

denuded hills and pseudo-alpine meadows one can see in Greece today.

This persistent model of a once-green-Greece seems to derive from Western Europeans and North Americans unconsciously perceiving the scenes they read in the ancient authors as taking place in an environment and climate much like their own—or accept uncritically the environmental destruction described by ancient moralizing essays like Plato's *Kritias*.[39] Take for example, the 17th century novelist J.J. Barthélemy, who sets the voyages of his hero Anacharsis in a Greece resembling the lush landscapes of Marie-Antoinette's France.[40] The view of a green paradise, destroyed by man has been around so long that it has become all pervasive, accepted, almost subconsciously evident. Consequently, most scholars of the ancient landscape, to a greater or lesser degree, include the destroyed environment in their arguments.[41]

But years of climatological and palynological analysis suggest that such a view is inaccurate.[42] As Oliver Rackham

[39] For a critique of these and other pseudo-historical environmental methodologies see O. Rackham, "The countryside: history and pseudo-history," *The Historian* 14 (1987): 13-17; "Ancient landscapes," in O. Murray and S. Price, eds., *The Greek City: From Homer to Alexander* (Oxford: Clarendon Press, 1990), 85-111; "Ecology and Pseudo-Ecology," in G. Shipley and J. Salmon, eds., *Human Landscapes in Classical Antiquity* (London and New York: Routledge, 1996), 16-43.

[40] J. J. Barthélemy, *Voyage du jeune Anacharsis en Grèce* (Paris: Debure, 1788).

[41] E.g. J. D. Hughes, "Deforestation, Erosion, and Forest Management in Ancient Greece and Rome," *Journal of Forest History* 26 (1982): 60-75; "How the Ancients Viewed Deforestation," *Journal of Field Archaeology* 10 (1983): 437-45; *Pan's Travail*; J. R. McNeil, *Mountains of the Mediterranean World*, and R. Meiggs, *Trees and Timber in the Ancient World* (Oxford: Clarendon Press, 1982). It is this long-standing perspective that underlies Hastead's assertions that the mountains in antiquity were too forested to yield acceptable pasture without extensive woodcutting and pasture-clearing

[42] E.g. H. E. Wright, "Vegetation History," in W. A. McDonald and G. Rapp, edd. *The Minnesota Messenia Expedition* (Minneapolis: University of Minnesota Press, 1972), 188-99; M. C. Sheehan, *The postglacial vegetational history of the Argolid Peninsula, Greece* (Diss., Univ Indiana, 1979); S. Bottema, "Pollen Analytical Investigations in Thessaly," *Palaeohistoria*, 21 (1979): 20-39; "Palynological Investigations on Crete," *Review Palaeobotany and Palynology*, 31 (1980): 193-217; "Palynological Investigations in Greece with Special Reference to Pollen as an Indicator of Human Activity,"

observes: "In general the Boeotian lanscape and vegetation appear to have changed less in 2,500 years than those of England in the last 1,000 or of New England in the last 180 years."[43] Environmental historians like Rackham have consistently demonstrated that the Greek climate and environment has not "degraded" since antiquity.[44] If anything, Greece has become *more* forested in the last hundred years because of the introduction of non-native species and aggressive fire and forest management practices. Consequently, climate-affected phenomena such as tree lines and alpine meadows were much the same or even more extensive in the ancient period than they are now. Consider again the example of Parnassos. Today, on Parnasos and its neighbors, the tree line is 2,400-2,500m; even the hardiest tree of the region, the Balkan Pine (*Pinus heldrechee*), can grow no higher than 2,400m because of the harsh temperatures at the upper elevations, while the more common Black Pine (*Pinus negra*) grows no higher than 1900m. As a result, from 2,500-3,000m alpine meadows consisting of tufted grasses dominate the upland landscape, meadows which heavy snow covers for over half of the year, resulting in a summer graze well-watered by snowmelt.[45] It seems that Semple was correct: the historical Greek environment has not changed substantially since antiquity.

While Halstead's assertions about the lack of pasture in ancient Greece have been questioned, his model of the intensively worked, agro-pastoralist farm has received much support. Stephen Hodkinson in particular was drawn to the new model and applied it directly to the subsistence and surplus economies of the Classical Greek polis.[46] Using

---

Palaeohistoria (1982): 251-89; and A. T. Grove and O. Rackham, *The nature of Mediterranean Europe.*

[43] O. Rackham, Observations on the historical ecology of Boeotia," 346.

[44] See the literature collected in O. Rackham, *Trees, wood and timber in Greek history* (Oxford: Leopard's Press, 2001); and A. T Grove and O. Rackham, *The nature of Mediterranean Europe.*

[45] McNeil, *Mountains of the Mediterranean World*, 18, 27.

[46] S. Hodkinson, "Animal Husbandry."

both literary and epigraphic evidence, he argued that intensive mixed farming, consisting of pastoral, arable, and arborial elements, was the desired production strategy for Classical period agriculture. Transhumance and other non-integrative strategies only became necessary when the resources near the farm became over-extended and could no longer support animals year-round. For Hodkinson, only very large scale operations would require separation of arable farming and animal production. But like Halstead, Hodkinson cautions that such specialized pastoralism would require a steady market or social demand to offset the effort and costs of such large-scale production. Hodkinson sees transhumance only as a last resort measure, unnecessary for the majority of animal producers.

Not so Jens Skydsgaard, who has vigorously critiqued the agro-pastorialist model. Skydsgaard returned to Semple and Georgoudi's transhumance thesis and argued that mobile strategies of moving herds from winter to summer pastures were much more evident in the literary and epigraphic sources than intensive fodder-cropping, especially among the strategies of the wealthy, large-scale producers, with whom the literary sources are most concerned. In addition, Skydsgaard called attention to the tendency of the ancient authors to draw a marked distinction between large-scale stockraising and arable agriculture.[47] In the ancient mind, Skydsgaard argues, farming and animal production is distinct and separate. He pointed out that there is no reason (nor much hard evidence) to conclude animal producers were systematically planting fodder crops and intensively rearing animals just because such techniques were known at the time. Skydsgaard warns that it is not enough to argue for the

---

[47] Skydsgaard cites the Attic orators by way of example, who discuss animals as productive units, complete with herdsmen, distinctly separate from the arable farm. See discussion below.

preference of fodder cropping over transhumance simply because it might yield the greatest economic returns.[48]

And so the matter stood for some time, with debate fossilized between the proponents of agro-pastoralism and transhumance. In their attempts to respond to their critics and focus on the minutiae of animal production, scholars of ancient Greek animal husbandry effectively isolated their subject from mainstream historical discussions of the ancient countryside and environment. The wider connections between city and countryside, between subsistence and surplus, between elites and non-elites, taking place in studies about the ancient landscape and economy were absent from the specialized agro-pastoralist/transhumance debate. Moreover, the agro-pastoralist/transhumance debate, with its generalized, uniform production rubrics, did little to increase understanding of ancient Greek animal husbandry itself, since by their general nature these broad models tended to overlook particulars such as regional variation and change in practice over time.

In 2003, Christophe Chandezon attempted to remedy this scattershot picture and offer a more regionally and chronologically sensitive approach by organizing the epigraphic evidence available for Greek stock-rearing in the eastern Mediterranean by region and date, from the late fifth century BCE to the late first century CE.[49] By assembling every published text concerned with animal husbandry in one place, and offering a critical translation of each, Chandezon created an invaluable database from which regional and temporal differences might be identified and measured. In the second half of the book, he offered explanations for some of the patterns revealed by the epigraphic evidence. In Western and Central Greece

[48] J. Skydsgaard, "Transhumance in Ancient Greece," in C. R. Whittaker, ed., *Pastoral Economies of Ancient Greece and Rome* (Cambridge Philological Society, Suppl. Vol. 41) (Cambridge: Cambridge University Press, 1988), 75-86. See also S. Isager and J. Skydsgaard, *Ancient Greek Agriculture: An Introduction* (London: Routlege, 1992), 99ff.

[49] C. Chandezon, *L'élevage en Grèce (fin Ve-fin Ier s. a.C.)*.

and the Peloponnese, for example, Chandezon observed
that pasture was fairly plentiful and generally situated
away from agricultural areas, very often in the uplands. In
the Aegean, and the drier eastern regions like the Argolid
and Attika, pasture was in shorter supply, leading to
greater integration with arable farming, which was later
accentuated by the need to intensify grain production in
the 4th and 3rd centuries. In essence, Chandezon
demonstrated that both the agro-pastoralist and
transhumance models were correct, but for different
regions, during different times, and because of different
environmental and societal pressures. Although the
political fragmentation of the Mediterranean landscape,
and the scarcity of good grazing land, encouraged large-
scale animal producers to move beyond their farm bases in
order to maintain large herds, that patchwork of
communities fiercely defending the integrity of their
territories and their agricultural resources created real
obstacles to long-distance mobility. Social constructs such
as political boundaries, sacred lands, taxation, royal policy,
and warfare all imposed limits on animal movement and
thus the scope and scale of animal production.

Although Chandezon's synthesis of the epigraphic
evidence greatly deepened our understanding of the
regional and temporal complexity of animal management
strategies and added some complexity to the agro-
pastoralist/transhumance debate, Chandezon did little to
bring the study of animal husbandry into mainstream
historical discussions. For the most part, he sidestepped the
agro-pastoralist/transhumance issue and even though he
identified the role of socio-political institutions, such as
territorial boundaries and religious restrictions, as
important variables shaping animal production, he does not
explain the wider context that underlay those choices.
Moreover, by focusing primarily on the epigraphic
evidence, which is weighted towards the 3rd-1st centuries
BCE, Chandezon's study is oriented more towards the
Hellenistic kingdoms and Roman dependencies than the
world of the Archaic and Classical Greek polis, from which

most of the literary sources derive. By ignoring the literary sources, Chandezon's study did not have the evidence to answer the fundamental question of "why animals?" And in order to move beyond the fossilized agro-pastoralist/transhumance debate and fully understand the context and regional complexity of animal husbandry for the Greeks we need to explain "why animals?" Just as Osborne, Garnsey, Sallares and others explained "why agriculture," we need to grapple with the social, political and economic choices and incentives that shaped large-scale animal production.

While the agro-pastoralist/transhumance debate was at its height, Hamish Forbes argued that understanding the social and political choices that underpinned animal production strategies was best way out of the current scholarly deadlock.[50] Forbes urged that attention be centered closely on the role of animals in the wealth-generating activities of the elite, rather than the purely subsistence-generating strategies of the masses, since the ancient literary and epigraphic sources focus more on the actions of the elite.[51] Drawing on his own ethnoarchaeological observations in ancient and modern Methana, and the work of others among the rural sites of the ancient Argolid,[52] Forbes demonstrated the extent to which the animal management strategies of the wealthy depended on a mosaic of variables such as terrain, fodder availability and market demand. Like Chandezon, Forbes

---

[50] Unfortunately, his suggestion has received little attention. H. A. Forbes, "The identification of pastoralist sites within the context of estate-based agriculture in ancient Greece: beyond the Transhumance versus Agro-pastoralism debate," ABSA 90 (1995): 325-338; C. Mee and H. A. Forbes, A Rough and Rocky Place: The Landscape and Settlement History of the Methana Peninsula, Greece (Liverpool: Liverpool University Press, 1997).

[51] Like Sallares (Ecology of Ancient Greece) and others, Forbes argues that the subsistence strategies of the masses had little or no real impact on the production of animals in antiquity.

[52] T. H. Van Andel and C, Runnels. Beyond the Acropolis. A Rural Greek Past (Stanford: Standford University Press, 1987); and M. H. Jameson, C. N. Runnels, and T. H. van Andel, edd., A Greek Countryside. The Southern Argolid from Prehistory to the Present Day (Stanford: Stanford University Press, 1994).

showed that both Hodkinson and Skydsgaard were correct: animal husbandry above the household, subsistence level seems to have been a mobile economic activity of the elite, at times fully integrated with arable and arborial activities, but also heavily dependent on uncultivated, marginal land, often located at some distance from the main agricultural base.

This book takes up Forbes' new direction. If the study of animal husbandry is to make any contribution to the wider history of Greece, specialists need to move on from isolated discussions of animal management strategies and embrace and explain the social (as well as economic) dynamics involved in large-scale animal production. In short, we need to explain why elites chose to raise animals in the first place and how such elite animal production activities affected Greek society at large. What follows is an attempt to unlock and contextualize pastoral politics. Chapter Two lays out the conceptual theme of the book by demonstrating that elite attitudes towards animal wealth were fairly uniform across Greece, inherently conservative in nature, and like many conceptions of elite behavior, slow to change over the Archaic and Classical periods. This conclusion, that the elite continued to view animal husbandry as an appropriate and desirable source of wealth, offers an adjustment to the well established social evolutionary model constructed by Anthony Snodgrass and made popular by Victor Davis Hanson, that arable farming replaced pastoral production at the beginning of the Archaic period.[53] Chapter Three re-opens the animal management debate and shows that animal management

[53] A. M. Snodgrass, *The Dark Age of Greece* (Edinburgh: Edinburgh University Press, 1971); *Archaic Greece: The Age of Experiment* (Berkeley and Los Angeles: University of California Press, 1980); and especially, *An Archaeology of Greece: The Present and Future Scope of a Discipline.* (Berkeley and Los Angeles: University of California Press). V. D. Hanson, *The Other Greeks.* This model has become so pervasive, it is accepted by most general textbooks. See, for example, S. B. Pomeroy, S. M. Burstein, W. Donlan, J. Roberts, *Ancient Greece: a political, social, and cultural history* (Oxford: Oxford University Press, 1999), 41-81. For a more cautious treatment see J. M. Hall, *A History of the Archaic Greek World ca. 1200-479* (Malden, MA: Blackwell, 2007), 61.

strategies should be treated as discrete regional and temporal phenomena, rather than pan-Hellenic, synchronic phenomena. Animal management strategies were highly tuned systems, linked to observable social and cultural behaviors, and thus the result of unique social, political and environmental pressures. Chapter Four shows the impact elite systems of animal husbandry had on Greek society at large by exploring the ways in which surplus, large-scale animal production could affect ordinary citizens through war, both sacred and secular. The primary motivation for many of the wars of the Archaic and Classical periods was an abiding need among the elite to secure access to scarce grazing resources. Non-elites simply were caught up in events. Finally, Chapter Five unites all of these themes by analyzing how the end products of elite animal husbandry, such as sacrificial displays, equestrian competitions, and feasts, offset the expenses involved in large-scale animal production by creating public goodwill, accentuating elite status, and creating real social and political power.

II

ANIMALS AS GENTLEMANLY WEALTH

Let me give you some idea of Odysseus' wealth.
On the mainland, twelve herds of cattle,
as many flocks of sheep, as many droves of pigs,
and as many scattered herds of goats,
all tended by hired labor or his own herdsmen;
while here in Ithaka eleven herds of goats,
graze up and down the coast,
with reliable men to look after them. [1]

Odyssey 14.99-104

As the first surviving works of Greek literature, the *Iliad* and *Odyssey* often serve as a benchmark, a point of departure, for developmental studies of ancient Greek attitudes and behavior. In part, this is because the eighth century BCE witnessed a significant rise in social complexity throughout many communities in Greece, both on the mainland and in Asia Minor, and in the course of this dramatic social upheaval cultural institutions were significantly modified. The most visible of these changes, the appearance of the Greek polis, hoplite warfare, and rural and urban sanctuaries, have been extensively studied. [2] As a result, because so many studies of early

---

[1] Translated by A. Burford, *Land and Labor in the Greek World* (Baltimore: Johns Hopkins University Press, 1993), 144.

[2] For a general discussion of the emergence of the polis see J. M. Hall, *A history of the archaic Greek world, ca. 1200-479 BCE* (Malden, MA : Blackwell, 2007). For more nuanced treatments see the essays collected in L. G. Mitchell and P. J. Rhodes, edd., *The Development of the polis in Archaic Greece*, (London: Routledge, 1997) and N. Fisher and H. van Wees, edd.,

Greek society have documented a general increase in population, site size, and institutional complexity, developmental models have come to dominate our understanding of early Greek history.[3] Consequently, many aspects of Greek society in the Archaic and later periods are viewed in terms of evolution, in terms of a progressive divergence from a Homeric, 8th century prototype.[4] Unfortunately, such emphasis on cultural

---

Archaic Greece: New Approaches and New Evidence (London: Duckworth, 1998). For early Greek warfare see A. M. Snodgrass, "The Hoplite Reform and History," *JHS* 85(1965): 110-22; H. van Wees, *Status Warriors: War, Violence and Society in Homer and History* (Amsterdam: J.C. Gieben, 1992); K. Raaflaub, "Soldiers, Citizens and the Evolution of the Early Greek *Polis*," in *The Development of the polis in Archaic Greece*, L. G. Mitchell and P. J. Rhodes, edd., (London: Routledge, 1997), 49-59; H. van Wees, "Greeks Bearing Arms. The state, the leisure class, and the display of weapons in archaic Greece," in *Archaic Greece: New Approaches and New Evidence*, N. Fisher and H. van Wees, edd., (London: Duckworth, 1998), 333-377; V. D. Hanson, *The Other Greeks* (Berkeley and Los Angeles: University of California Press, 1999), 219-286; V. D Hanson, "Hoplite Battle as Ancient Greek Warfare. When, where, and why?," in *War and Violence in Ancient Greece*, H. van Wees, ed.,(London: Duckworth, 2000), 201-232; H. van Wees, "The Development of the Hoplite Phalanx. Iconography and reality in the seventh century," in H. van Wees, ed., *War and Violence in Ancient Greece* (London: Duckworth, 2000), 125-166; and H. van Wees, *Greek Warfare. Myths and Realities* (London: Duckworth, 2004). For religious sites see F. de Polignac, *La Naissance de la cité greque* (Paris: Edition La Découverte, 1985), C. Morgan, *Athletes and Oracles* (Cambridge: Cambridge University Press, 1990); C. Sourvinou-Inwood, *"Reading" Greek culture : texts and images, rituals and myths* (Oxford: Oxford University Press, 1991); N. Marinatos and R. Hågg, edd., *Greek Sanctuaries. New Approaches* (London: Routledge, 1993); S. Alcock and R. Osborne, *Placing the Gods. Sanctuaries and Sacred Space in Ancient Greece* (Oxford: The Clarendon Press, 1994); C. Morgan, *Early Greek States Beyond the Polis* (London: Routledge, 2003).

[3] S. Morris, *Daidalos and the Origins of Greek Art* (Princeton: Princeton University Press, 1992), and C. Sourvinou-Inwood, "Early sanctuaries, the eighth century and ritual space. Fragments of a discourse," in N. Marinatos and R. Hagg, edd., *Greek Sanctuaries* (London: Routledge, 1993), 1-17, do not accept this explosion of activity in the eighth century, but rather a slowly changing society, adopting the characteristics of the polis over a long period of time.

[4] Éduard Will in his 1957 work, "Aux origines du régime foncier grec. Homère, Hésiode et l'arrière-plan Mycénien," *Révue des Études Anciennes* 59 (1957): 5-50, pioneered this social/political evolutionary model, which has become widely accepted in Greek historical scholarship. See A. Edwards, *Hesiod's Ascra* (Berkeley and Los Angeles: University of California Press, 2004), for discussion and critique.

progression tends to obscure the more conservative aspects of Greek social behavior and even disregard Archaic and Classical reverence for Homeric archetypes as models for gentlemanly conduct.[5]

The peoples of the ancient Mediterranean world were extremely conservative in their views about gentlemanly activities, especially wealth and wealth production.[6] These values and customs were consciously conservative and changed very slowly, if at all. In order to preserve their coveted status as elites, the powerful were particularly interested in identifying, acquiring, and praising "proper" sources of gentlemanly wealth. Appearances were everything and throughout their relatively long cultural histories the Greek and Roman elites thought and wrote a great deal about what constituted suitable forms of wealth, and what did not. Not only did these largely aristocratic writers perpetuate the same conclusions—that proper wealth was gained from the land and the products of agriculture—but they even weighted their arguments with social and moral biases, asserting that the most noble and honorable men, the men free from the mundane cares of subsistence and survival, were those who eschewed industry and commerce and engaged only in the purest of agricultural pursuits. The epics of Homer, for example, stereotyped those who procured wealth through trade as "rascally" and "grasping" but portrayed herd owners and agriculturalists as "wealthy" and "virtuous."[7] In a similar

[5] A. W. H. Adkins, *Merit and Responsibility: A study in Greek Values* (Oxford: Oxford University Press, 1960); *Moral Vales and Political Behaviour in Ancient Greece* (New York: W. W. Norton, 1972); and W. Donlan, *The Aristocratic Ideal in Ancient Greece* (Lawrence, Kansas: Coronado Press, 1980).

[6] Ian Morris, "Archaeology and Greek History," in N. Fisher and H. van Wees, edd., *Archaic Greece: New Approaches and New Evidence* (London: Duckworth, 1998), 1-91, argues that in many areas, especially the Peloponnese, Crete, and Northern and Western Greece, elite displays of wealth and consequently elite behavior, remained largely unchanged down through early Classical times.

[7] *Od.* 14.288-9; 15. 415-17. Cf. *Od.* 14.99-104. Ian Morris, "The Use and Abuse of Homer," *Classical Antiquity* 5 (1985): 115-125, argues that the Homeric epics were an attempt by the eighth century elite to justify and entrench a class ideology.

fashion, the 6th century BCE Megarian poet Theognis, who was known for his hatred of Megara's rising wealthy, contrasted the despicable "cargo-carriers" and the agricultural "best men."[8] Centuries later, Aristotle praised the good, traditional, agricultural and pastoral forms of wealth-getting as gateways to the "good life," while at the same time condemning the wrong, self-serving, commercial wealth-getting of the grasping mercantilists, who did not even seek the good life but only more coin.[9] But even Aristotle admitted that commercial gain was not in itself inherently evil and could be redeemed, so long as it was put to a good use—that is, invested in agriculture.[10] It seems that Aristotle was willing to admit new men into the gentlemen's club, provided that they conformed to proper standards of behavior and divested themselves of improper forms of wealth.

For these elite writers and their elite audiences, reaping nature's bounty was *by nature* superior to trading or manufacturing goods and therefore the proper pursuit for the socially superior man. Notice the terms in which Cicero, a former outsider to the Roman nobility and thus eager to exhibit all the proper attitudes, expresses the aristocratic perspective: "of all the ways of acquiring wealth, none are better than agriculture, none more pleasant and fruitful, none more fitting to a free man."[11] If possible, the Greek elite expressed this agrarian bias even more forcefully than the Romans: "Whoever said that agriculture is the mother and nurse of all the other

---

[8] Theognis 679ff.

[9] Arist. *Pol.* 1257a-1258a.

[10] This is not all that surprising, since some of the most notable elites from Classical Athens built their fortunes in non-agricultural ways. Nikias, the famous Athenian general and his father had made much of their wealth from the hide-tanning industry, which they then invested in estates and farms. J. K. Davies, *Athenian Propertied Families 600-300 B.C.* (Oxford: The Clarendon Press, 1971), 403, and S. Hodkinson, "Politics as a determinant of pastoralism: the case of southern Greece," *Rivista di studi liguri* 16 (1992): 139-64. After the restitution of democracy at Athens in the fourth century BCE, a certain upwardly-mobile Nausikydes invested the profits from his slave-run milling business into large herds of cattle and pigs (Xen. *Mem.* 2.7.6).

[11] *De Off.* 1.151.

disciplines spoke nobly indeed. For when agriculture goes well, all the other skills also flourish, but when the earth is compelled to lie barren, the other disciplines almost cease to exist, on sea, as well as on earth."[12] But agriculture is a wide subject. The ancient Greek gentleman could be much more discerning, even snobbish, in distinguishing between the most proper forms of agricultural production. There seems to have existed a hierarchy of agricultural pursuits, with animal production slightly more proper, slightly more suitable for the gentleman than cereal and arborial agriculture. Aristotle explains this view:

> The most useful types of wealth production are first knowledge concerning beasts, which are most profitable where and how, that is knowledge concerning production of horses, or of cattle, or of sheep, and likewise of the remaining animals. For one must be expert as to which animals are most profitable compared to each other, and also which are most profitable on what sorts of land, for some thrive on different lands. Then knowledge of agriculture...and finally of bee-keeping and of other animals, feathered and finned which can bring wealth. These are the types and primary parts of wealth-getting in the most proper sense.[13]

While Aristotle acknowledges agriculture as a worthy pursuit, a suitable form of gaining wealth, he gives the production of animals priority, and even implies that there is a hierarchy among types of animals, with horses being most prestigious, followed by cattle, sheep and other beasts.

Several factors may account for the priority of animal husbandry over agriculture, as well as the priority of large animals like horses and cattle over small animals like sheep and swine. Large animals like horses and cattle are difficult and expensive to produce, even in small numbers. Small animals are much easier to maintain; sheep and goats can eat otherwise useless scrub brush, and swine can be fed

[12] Xen. *Oec.* 5.17.
[13] Arist. *Pol.* 1258b.12-21.

table scraps and even human waste. Even small numbers of horses and cattle require great amounts of specially grown feed, open pasture, and personal care by herdsmen. Because of these investments in upkeep and land, animals, especially horses and cattle make a large public impression, much more of an impression than huge tracts of grain or spreading orchards. Even in our modern industrial society, a Grand Champion bull or a Triple Crown winning race horse makes a powerful statement: the initial purchase and maintenance cost of breeding stock, staff, feed, pasture, shelter and other daily concerns are implied, if not fully understood by even the most sheltered suburbanite. In addition, because of the inherent cost, top-quality (and large-scale) animal husbandry is a naturally selective pursuit, an activity that only a very few could afford. In the ancient Mediterranean world, only the exceptionally wealthy could provide the fodder and manpower resources necessary to raise herds of animals, in such a dry region, where land for grazing was scarce and consequently would have to be removed from food production and devoted specifically to pasture or animal fodder, perhaps even irrigated with equally scarce water supplies. To put it another way, producing animals was the most fitting pursuit for the Greek gentleman simply because the high investment costs excluded almost everyone else.[14] Consequently, animals, particularly large animals such as cattle and horses, became naturally exclusive symbols of

---

[14] And during the late Archaic and early Classical period, other traditional "elite" pursuits such as orientalizing art were attracting competition from non-elites as well as unwanted political connotations. See I. Morris, "The Art of Citizenship," in S. Langdon, ed., *New Light on a Dark Age* (Colombia, MO: University of Missouri Press, 1997), 9-43; L. Foxhall, "Cargoes of the Heart's Desire. The character of trade in the archaic Mediterranean world," in N. Fisher and H van Wees, edd., *Archaic Greece: New Approaches and New Evidence* (London: Duckworth, 1998), 295-309; and S. Hodkinson, "Lakonian Artistic Production and the Problem of Spartan Austerity," in N. Fisher and H van Wees, edd., *Archaic Greece: New Approaches and New Evidence* (London: Duckworth, 1998), 93-117.

wealth. Herds and flocks, by their very nature, best displayed the eliteness of their owner.[15]

From Homeric times down to the time of Aristotle the literary sources celebrate the fact that animals constituted the proper forms of gentlemanly wealth. And yet, such an assertion about stagnated social values, about lack of social change during the dynamic Archaic and Classical periods, seems to contradict not only our current understanding of the development of both agriculture and animal husbandry in Greece, but even the accepted view of evolving aristocratic priorities and values in the age of the emerging polis. The prevailing developmental model of animal husbandry, for example, asserts that "the peak of livestock farming occurred in the earlier part of the Iron Age, and...a falling off occurred with the approach of the historical period."[16] According to Anthony Snodgrass, the Early Iron Age communities diverted a greater part of their resources to the pastoral sector than before (or after) because the collapse of the Mycenaean palace economy disrupted the existing system of landholding and resulted in the abandonment of agricultural land and sedentary agriculture in general. As the refugees from the Mycenaean collapse

[15] J. K. Campbell in his work among the Sarakatsani, a group of modern pastoralists in central Greece, explains how animals come to form a critical element in social reputations. Taken together, the numbers and quality of animals form the necessary material for high prestige: they imply self-sufficiency and ability, both economically in terms of subsistence and marketable surplus, and technically in terms of manpower and resources. A man of flocks draws to himself a number of dependent individuals as herdsmen, and because of these dependent relationships, a man of flocks carries weight in the formal setting of the council as well as the more informal setting of the marketplace. These men because of their visible success in raising animals are given respect and loyalty. J. K. Campbell, *Honour, Family, and Patronage: a study of institutions and moral values in a Greek mountain community* (Oxford: The Clarendon Press,1964).

[16] A. M. Snodgrass, *An Archaeology of Greece: The Present and Future Scope of a Discipline* (Berkeley and Los Angeles: University of California Press, 1987), 207. See also *The Dark Age of Greece* (Edinburgh: Edinburgh University Press, 1971), 180-181; and *Archaic Greece: The Age of Experiment* (Berkeley and Los Angeles: University of California Press, 1980). Such models are not unique to Late Bronze and Early Iron Age Greece. See W. G. Dever, *Who were the Early Israelites and Where did they come from?* (Grand Rapids, MI: Eerdmans, 2003), for a survey and critique of models of pastoral reversion after the collapse of the Bronze Age civilizations of the Eastern Mediterranean.

fled the areas of destruction, they were forced to adopt different systems of food production, more compatible with their new, mobile lifestyle. This subsistence shift, brought about by necessity, became firmly established and remained a significant factor in Early Iron Age society for several hundred years, even after people began to form sedentary communities and once again practice arable and arborial agriculture. Only the return to sedentism, marked by an increase in population and the pressures of the emerging polis, brought an end to specialized pastoral production.

To support this theory of Early Iron Age pastoralism, Snodgrass noted the appearance of the following: (1) land more available after the collapse of the Mycenaean palace civilization due to massive depopulation; (2) an increase in animal bones from the excavated site of Nichoria, suggesting a major shift to beef-ranching; (3) the preservation for centuries of the names of long-abandoned sites, suggesting a shift in local settlement behavior but not inhabitants, as the newly-mobile pastoralists tended herds in areas where they had once lived; (4) the increase in handmade pottery, suggesting mobile potters; (5) circular huts replacing more permanent stone and timber structures, suggesting short-term settlement; (6) apsidal house plans, which recall similar structures used among mobile pastoralists from other societies and regions; (7) the appearance of Hesiod's poems at the end of the Dark Age to teach the new practice of agriculture to an audience unfamiliar with it; and (8) Early Iron Age sanctuary deposits of animal figurines, which decline drastically after 800 BCE, suggesting a decline in the overall importance of animals. Although Snodgrass admits that these examples are weak on their own and do not by themselves make a compelling case for specialized pastoralism, he argues that together they have a certain force that must be explained.

Soon after Snodgrass published this thesis, John Cherry offered a thorough critique.[17] Cherry observed that (1)

---

[17] J. Cherry, "Pastoralism and the Role of Animals in the Pre- and Protohistoric Economies of the Aegean," in C. R. Whittaker, ed., *Pastoral*

handmade pottery and apsidal houses are not found only among pastoralists; (2) Hesiod's poems also include advice on caring for animals, such as when to castrate, when to shear, and seem a poor introduction to sedentary agriculture for those unfamiliar to it; and (3) there exists no archaeological or literary evidence during the Greek Dark Age for any mode of subsistence *or* market-oriented production based primarily on pastoralism. In fact, Cherry examined the archaeological and historical records from Neolithic to modern times for examples of any types of pastoral production and concluded that specialized pastoralism, in which humans relied mostly or exclusively on animal resources, existed historically only when an elaborate system of exchange for pastoral products was in place between pastoralists and agriculturalists, a system for which there is no evidence during the Greek Dark Age.[18] Moreover, Cherry argued that Snodgrass' entire pastoral theory hinged on a developmental model that assumed sedentism was a desired condition, and that a "reversion" to mobile pastoralism will occur only when socio-political or economic conditions leave no other alternatives. According to Cherry, Snodgrass' model of the civilized, agricultural Mycenaean world devolving into a backward, pastoral Dark Age society of simple herders is methodologically flawed, based more on assumptions about social evolution than sound archaeological or textual evidence.

Even after Cherry's criticisms, however, many scholars were reluctant to lay aside the model of Dark Age pastoralism.[19] As a result, Lin Foxhall returned once more

---

Economies of Ancient Greece and Rome (Cambridge Philological Society, Suppl. Vol. 41) (Cambridge: Cambridge University Press, 1988), 196-209.

[18] See L. Nixon and S. Price,"The Diachronic Analysis of Pastoralism through Comparative Variables," *ABSA* 96 (2001): 395-424, for the necessary connections between pastoralism and the market economy.

[19] For example, C. Morgan, *Athletes and Oracles* (Cambridge: Cambridge University Press 1990). This model has become so pervasive that it is has crept into most Greek history textbooks. See, for example, S. B. Pomeroy, S. M. Burstein, W. Donlan, J. Roberts, *Ancient Greece: a political, social, and cultural history* (Oxford: Oxford University Press, 1999), 41-81. For a more

to the evidence and re-examined Snodgrass' archaeological conclusions.[20] The countryside of the Early Iron Age was not as barren and depopulated as Snodgrass had proposed; recent survey work has revealed a patchwork of sites otherwise unknown. While the framework of Mycenaean society changed dramatically in the Late Bronze Age, Foxhall observed that contrary to Snodgrass' assertions, many of the individual components of Mycenaean society changed very little, and that the extent to which the collapse of the palace system affected people depended largely on their proximity to a palace center and their degree of integration into palace economy and society. Consequently, Foxhall sees no "complete breakdown" of social order and reversion to a backward tribal or pastoral society. Indeed, she argues that many communities of Early Iron Age Greece, such as Nichoria, Tiryns, and Lefkandi, retained nucleated settlements with complex "state-like" political configurations, active elites, and an unabated, unchanged exploitation of the surrounding territory. In the end, only the production of certain luxury items such as perfumes, textiles and monumental art and architecture ceased with the Mycenaean collapse. The production of food continued much as it had always done, no more or less pastoral than before.

Despite the work of Cherry and Foxhall, a developmental view of Early Iron Age pastoral production still persists. Victor Davis Hanson, in his controversial book *The Other Greeks. The Family Farm and the Agrarian Roots of Western Civilization,* has given new life to the theory of Dark Age pastoralism by offering some social and cultural reasons for the apparent behavioral shift from predominantly pastoral to predominantly agricultural

---

cautious treatment see J. M. Hall, *A History of the Archaic Greek World ca. 1200-479*, 61.

[20] L. Foxhall, "Bronze to Iron: Agricultural Systems and Political Structures in Late Bronze Age and Early Iron Age Greece," *ABSA* 90 (1995): 239-250.

production.[21] In his original argument, Snodgrass had suggested that the agricultural focus of Hesiod, who was writing around 700 BCE, presaged the end of the pastoral regime. At that time, the Greeks abandoned specialized pastoralism and returned to sedentary agriculture. Hanson sought to explain the reasons behind this subsistence shift, postulating that at some time after the collapse of the Mycenaean civilization, although the elite were primarily engaged in animal production, there was a gradual movement of non-elites towards small-scale diversified agriculture. Thus, during the later years of the Greek Dark Age, before the time of Hesiod, those men who did not own herds or agricultural bottomland of their own moved out onto the hillsides and began to carve out small farms and raise their own food. Their ranks swelled as more men became dissatisfied with the cattle-rearing aristocrats and their total control of the fertile bottomland pastures. For Hanson, access to farmland and household self-sufficiency provided the primary motivation for this shift in agricultural practice. The small farms, eked out on marginal lands and intensively worked by fiercely independent owners, provided the physical and ideological base for the "Archaic agricultural revolution," and indeed for the creation of the Greek polis and hoplite warfare as well. By the beginning of the eighth century, most Greeks had returned to sedentary agriculture, leaving only an increasingly beleaguered elite engaged in animal production. According to Hanson, this tension over access to land between the stockrearing aristocrat and hoplite farmer drove the major social and political achievements of the Archaic period, and resulted in the gradual replacement of the pastoral elites everywhere in southern Greece.

Hanson's picture of the "Archaic agricultural revolution" is flawed for several reasons. First, the vision

[21] V. D. Hanson, *The Other Greeks. The Family Farm and the Agrarian Roots of Western Civilization* (Berkeley and Los Angeles: University of California Press, 1999). For a more cautious use of Snodgrass' pastoral model see C. Thomas and C. Conant, *Citadel to City State. The transformation of Greece, 1200-700 B.C.E.* (Bloomington, ID: Indiana University Press, 1999), 32-59.

of the individualistic, middle-class hoplite "homesteader" is a particularly modern, American notion, and has little place in any European society, ancient or modern.[22] Second, and more importantly, Hanson's theory of independent farmers, living on isolated homesteads, cultivating marginal land contradicts the available archaeological evidence. Archaeological surveys of the ancient Greek countryside do not support any ninth and eighth century movement of disgruntled farmers out to marginal lands. In fact, this type of rural habitation seems to have been a much later phenomenon, known only in the late fifth and fourth centuries BCE, and then only in a few regions of Central Greece such as Attika and the Argolid.[23] And yet, despite these methodological flaws, Hanson's reassessment of the Snodgrass thesis makes important observations about elite agricultural behavior during the Early Iron Age: (1) Dark Age animal production was under the control of an elite (an assertion with which even Hanson's critics agree); and (2) this pastoral elite continued to produce animals, even after the rest of society had abandoned pastoralism and returned to sedentary agriculture. In the end, Hanson's study is useful for understanding elite conceptions of wealth because it demonstrates that elite attitudes towards agriculture and animal husbandry changed little, even though many other aspects of Greek society were experiencing great pressure.[24]

---

[22] See L. Foxhall, ""Greek Agrariansim.' Review of Victor Davis Hanson, *The Other Greeks*, " *Classical Review* 48.2 (1998): 390-1.

[23] J. L. Bintliff and A. M. Snodgrass, "The Cambridge-Bradford Boeotian Expedition: the first four years," *Journal of Field Archaeology* 12 (1985): 123-161; H. Lohman, *Atene: Forschungen zu Siedlungs- und Wirtschafts-structur des klassischen Attik, I-II* (Köln: 1993); M. H. Jameson, C. N. Runnels, and T. van Andel, *A Greek countryside: the Southern Argolid from Prehistory to the present day* (Stanford: Stanford University Press, 1994); and C. Mee and H. A. Forbes, *A Rough and Rocky Place: The Landscape and Settlement History of the Methana Peninsula, Greece* (Liverpool: Liverpool University Press, 1997).

[24] Indeed, the fact that Aristotle, near the end of the polis period agues that "wealth consists of slaves, herds and money," suggests that elites' attitudes towards large-scale animal production had changed little during the Archaic and Classical periods. Arist. *Pol.* 1267b.10.

Beginning with the *Iliad* and *Odyssey*, we can see that herds of cattle and sheep played an important role for the ancient Greek elite.[25] Both Snodgrass and Hanson have used the prevalence of animals and animal feasts to argue that the Homeric household and subsistence economies were more pastoral than those of later Greeks.[26] And while they are correct in observing that animals had a central role in Homeric systems of wealth and prestige, perhaps they have overemphasized the general importance of pastoralism in relation to other subsistence activities. The epics are equally full of descriptions of cereal and arborial cultivation, and in all likelihood, Homeric subsistence strategies differed little from later Greek practice and were based firmly on arable and arborial farming.[27] The typical elite estate resembled that of Diomedes' father, "rich in grain land and orchards and many sheep."[28] Animal production, while an important activity for the elite heroes, was not their sole, or even primary subsistence activity.

So why did Diomedes' father have so many sheep, and why do the Homeric poems showcase animals, if they

[25] The *Iliad and Odyssey* have seemed to many scholars to describe a coherent, self-contained society. As Finley argued some time ago, the culture portrayed in the epics appeared to be a discrete institutional unit, accurately reflecting the society of the poet at the time of writing. M. I. Finley, "Homer and Mycenae: Property and Tenure," *Historia* 6 (1957): 133-59; *The World of Odysseus* (London: Chatto and Windus, 1978). Perhaps Ian Morris put it best when he observed that the epics represent the state of Greek society in the second half of the eighth century, with only occasional gaps and exaggerations. I. Morris,"The Use and Abuse of Homer," *Classical Antiquity* 5 (1985): 119. For similar conclusions and a recent discussion of the problems See K. A. Raaflaub, "A Historian's Headache. How to read 'Homeric Society'?," in N. Fisher and H van Wees, edd., *Archaic Greece: New Approaches and New Evidence* (London: Duckworth, 1998), 169-193.

[26] So, too, does W. Richter, *Die Landwirtschaft im homerischen Zeitalter. Mit einem Beitrag: Landwirtschaftlische Geräte, von Wolfgang Schiering* (Archaeologia Homerica II, H) (Göttingen, 1968).

[27] E.g. *Il.* 12.310-20; 14.121-4; *Od.* 7.110-32; 16.139-45. So argue J. Cherry, "Pastoralism and the Role of Animals in the Pre- and Protohistoric Economies of the Aegean"; L. Foxhall, "Bronze to Iron: Agricultural Systems and Political Structures in Late Bronze Age and Early Iron Age Greece"; and P. Halstead, "Pastoralism or household herding? Problems of scale and specialization in early Greek animal husbandry," *World Archaeology* 28 (1996): 20-42.

[28] *Il.* 14.124.

played little role in daily subsistence? Values. In the end, it all comes down to elite values. The poet gave animals priority because they were socially relevant to his audience. The Homeric heroes call attention to their herding activities because their reputations, their eliteness was measured by the quality and quantity of their animals. The distribution of animals in the form of gifts, or even as meat distributed at feasts, helped to maintain the complex reciprocal networks of elite society that were essential to the success of the aristocratic household.[29] Indeed, Homer often speaks of the importance of animals in creating and maintaining a hero's reputation. In Book Two of the *Odyssey*, for example, Telemachos laments the fact that his mother's suitors are destroying his source of livelihood, feasting in Odysseus' house, slaughtering his cattle, sheep, goats, and swine.[30] The slaughter is so upsetting to Telemachos because these animals represent the raw materials that he needs to demonstrate and prove his aristocratic status. Without the animals from his father's herds, Telemachos will be unable to hold feasts of his own and give expensive animal gifts, and if he wishes to become a fully functioning member of Homeric aristocratic society, he will need to hold those feasts and give those cattle.

Epithets such as "rich in flocks" and "rich in herds," borne by the geographical locations, groups of people and prominent individuals, attest to the close connections between animals and aristocratic reputation in the Homeric world.[31] One of the most explicit of these animal epithets is "rich in flocks," as in, "Phorbas, rich in flocks, whom

[29] W. Donlan, "Reciprocities in Homer," *The Classical World* 75 (1982): 137-75; "The Relations of Power in the Pre-State and Early State Polities," in L. G. Mitchell and P. J. Rhodes, edd., *The Development of the Polis in Archaic Greece* (London: Routledge, 1997), 39-48; and S. Hodkinson, "Imperialist Democracy and Market-Oriented Pastoral Production in Classical Athens," *Anthropolozoologica* 16 (1992): 53-61, have discussed the significance of animals at feasts or as gifts between peers. G. Herman, *Ritualized Friendship and the Greek City* (Cambridge: Cambridge University Press, 1987), argues that animals never seem to lose this role among the Greek elite.

[30] *Od.* 2.48ff.

[31] *Il.* 2.106, 605, 705; 9.154, 296; 14. 490, 16.417; *Od.* 11.257, 15.226.

Hermes [the god of flocks] gave wealth."[32] "Rich in crops," by contrast, appears only once in the Homeric corpus.[33] Possessing numerous animals is significant in this society and important enough to serve as a class-defining moniker. A cultural insider would decode Phorbas' epithet "rich in flocks" as implying that its owner was a gentleman of wealth, who has all of the requisite fodder, equipment—such as pens and barns—as well as the necessary labor to live the aristocratic lifestyle and thus to participate in proper elite behavior such as giving feasts and gifts. In short, the man rich in flocks has the proper sort of wealth with which to reciprocate as a gentleman.

In addition to signifying elite class status, abundance of animals is also used by the epic poet to convey the unparalleled and truly impressive quantity of resources owned by the very wealthy, the *basileis*. In Book Four of the *Iliad*, for example, the poet compares the clamoring, pushing warriors in the Trojan ranks to the many thousands of sheep jockeying in their pens, waiting to be milked on the estate of an extremely wealthy man. A more powerful example comes from the famous catalogue of Odysseus' wealth in book fourteen:

> Let me give you some idea of Odysseus' wealth.
> On the mainland, twelve herds of cattle,
> as many flocks of sheep, as many droves of pigs,
> and as many scattered herds of goats,
> all tended by hired labor or his own herdsmen;
> while here in Ithaka eleven herds of goats,
> graze up and down the coast,
> with reliable men to look after them.[34]

Here, a description of Odysseus' animal resources serves to acquaint the reader (listener) with Odysseus' aristocratic pedigree, his position as a *basileus*. In an effort to prove the stature and ability of his master to the foreigner, who is himself the disguised Odysseus, Eumaios

---

[32] *Il*. 14.490.
[33] *Il*. 5.613.
[34] *Od*. 14.99-104.

boasts about his master's pastoral wealth. In essence, Eumaios is using a catalogue of animal wealth to prove to the foreigner that despite what he has heard, and despite his master's long absence, Odysseus is a proper *basileus* and no ordinary mortal.

Apart from signifying elite status, cattle may have fulfilled an even more complex role in Homeric society, as measures of value in gift exchanges between the gentlemen. In a much analyzed passage, the Trojan hero Glaukos trades his gold armor, which the poet describes as "worth" 100 cattle, for the bronze arms of Diomedes, worth only nine.[35] Although this unique and troubling exchange has convinced some scholars that cattle served as an official standard of currency in Homeric society, the extent to which the economy was based on a "cattle-standard" is unknown and probably unknowable.[36] But all the same, some value hierarchy is at work. On some level, in a non-monetary society like Homeric Greece, a measure of value is necessary, if only so that a hero may measure the quality of a gift. Without some rubric in place the hero could not reciprocate with an equal or better gift. It is not surprising then, in a world where animals are so highly esteemed that an ox or bull would serve as a relative measure of value. Indeed, the fact that the exchange between Glaukos and Diomedes is a mutual but unequal gift underscores the important roles of animals in defining and maintaining social hierarchies and dependencies.

Like his Homeric forebears, the gentleman of the emerging polis continued to hold animal wealth in high esteem. The Homeric Hymn to Earth, for example, observes that "the man whom Earth best favors has fields abounding in herds, and both great material prosperity and wealth follow him."[37] It is significant that Earth, who is perhaps more naturally connected with arable agriculture, gives a man herds by which his prosperity becomes visible

[35] *Il.* 6.234.

[36] Richter, *Die Landwirtschaft im homerischen Zeitalter*, and Finley, *The World of Odysseus*.

[37] HH 30.10.

and his wealth can be measured. Here, perhaps for the first time, we see the hierarchy described by Aristotle centuries later, that the most useful type of wealth that the land can yield is animals.[38] Hesiod echoes the close relationship between animals, land, and the best forms of wealth. In the *Works and Days*, the poet admonishes his wastrel brother Perses that only "from work do men become well-flocked and wealthy."[39] Hesiod recognized that being well-flocked was difficult and thus in practice attainable by few men. In fact, he notes that only the most noble men, who have been blessed by the gods are well-flocked. It is significant that in the *Works and Days* only the blessed golden race, the best of all men, "dwell at ease, rich in flocks, and loved by the holy gods."[40] For Hesiod, having animal wealth was a defining characteristic of none but the most superior of men.

And such pastoral wealth seems to have imbued in these superior men a certain class consciousness and social power. Take, for example, the elite of Megara, who lost control over their city because they lost control of their animal wealth. Around 640 BCE Theagenes, a would-be tyrant, took control of Megara only after he gained control over the herds of the wealthy; Theagenes had tried to take the city before, without seizing the herds, and had failed.[41] In losing their animals it seems that the Megarian elite had lost an important link to their social power and were no longer socially powerful enough to resist Theagenes. This situation is strikingly similar to the one in which Telemachos found himself in the *Odyssey*. As long as the

[38] Arist. *Pol.* 1257a-1258a. See above discussion.

[39] WD 307.

[40] WD 307. Such statements by Hesiod in the *Works and Days* about animals and their role in the wealth production strategies of the best men show that both Snodgrass, *Archaeology of Greece*, and Hanson, *The Other Greeks*, exaggerate Hesiod's agricultural focus. No longer can we assert that Hesiod wrote this didactic poem as a guide to agricultural life for those who were unaccustomed to it. Hesiod was interested in both agricultural *and* pastoral issues; the poet himself was tending sheep on Helicon when the muses visited him and inspired him to compose the *Theogony*. See Edwards, *Hesiod's Ascra*.

[41] Arist. *Pol.* 1305a.28.

suitors consumed his herds, Telemachos was powerless. He could not access the symbols of a hero's kingly stature and reputation. Once Theagenes gained control of the ruling elite's herds and began to slaughter them, he publicly destroyed the physical representations of elite power, the symbols of their superiority. Control over the Megarian herds implied control over grazing land, over dependent shepherds, and over sacrificial gifts to gods and men. This control reinforced eliteness among the rulers of Megara and gave them dominance in important social, political and economic relationships within Megarian society.[42] By killing the herds, Theagenes killed the Megarian oligarchs' eliteness.

Despite the dramatic social changes wrought in the Archaic Greek communities by external war, internal *stasis*, and even tyrants, elite conceptions of animal wealth as marks of best men became even more entrenched during the late Archaic and Classical periods. Perhaps because of the social mobility made possible by the pluralistic political systems that tyrants like Theagenes and Peisistratos of Athens helped put into place, elite conceptions of proper wealth became even more entrenched. As discussed earlier, animals by their very nature are expensive and require a great deal of effort to maintain and breed. In the political reforms of the late Archaic, the newly wealthy had been given some political rights, but in places like Athens they were simply not given the landed resources necessary to raise large, impressive animals like horses and cattle. In communities like Athens, large animals (and even large herds of small animals) could still be the preserve of the old, landed families.[43] The epinikian works of the Theban

---

[42] H. van Wees, "Megara's Mafiosi: Timocracy and Violence in Theognis," in R. Brock and S. Hodkinson, *Alternatives to Athens. Varieties of Political Organization and Community in Ancient Greece* (Oxford: Oxford University Press, 2001), 52-67, suggests that control over status symbols and social relationships played a pivotal role in Archaic Megara.

[43] In the *Clouds*, Pheidippides, a young would-be aristocrat, was so obsessed with horses that he dreamed of them at night and squandered the

poet Pindar illustrate the role animals played in traditional gentlemanly identity at this time. In *Pythian* 4.148-150, for example, the poet recalls how the sheep and cattle of Pelias exhibited his wealth and maintained his house's reputation. And in *Olympian* 10.88-90 Pindar explains further the important status linkages between animal wealth and family identity: "wealth handed over to another sheep owner is most hateful for one dying without an heir." What seems especially hateful is that the heirless sheep owner cannot pass his status on to a son or close male relative and therefore has to give an outsider, perhaps even a rival, the symbols of his family's social prominence.

The court speeches of the later Classical period underscore this theme that animals create and maintain a man's social position. In the later fourth century, Philip of Macedon gave several Athenian ambassadors large numbers of cattle, sheep and horses in return for their continued support of his interests at Athens. Of course, this gift roused the ire of Demosthenes, who was highly critical of Philip and his expansionist policies.[44] Demosthenes contends that a certain Euthykrates "was keeping a large herd of cattle for which he had paid nothing to anyone, while another man returned home with a flock of sheep, and another with a herd of horses, and the masses (whose interests were being endangered) instead of becoming angry and demanding the punishment of the traitors, stared at them, envied them, honored them, and considered them true men."[45] What seems to have upset Demosthenes the most is that Euthykrates and his fellows had changed social status by means of Philip's gifts. Euthykrates and the others have become gentlemen, "true

---

family fortune in an unsuccessful attempt to acquire a winning team of racers (*Nubes* 21-22).

[44] See R. Sealey, *Demosthenes and his time: a study in defeat* (Oxford: Oxford University Press, 1993), and I. Worthington, ed., *Demosthenes: statesman and orator* (London: Routledge, 2000), for further discussion.

[45] 19.265.

men," and because of their new animal wealth have inspired both envy and honor in their fellow Athenians. Such envy and honor had previously belonged to Demosthenes and his peers and this new competition from men he considers his social inferiors has upset Demosthenes greatly.

While Demosthenes' anger is perhaps understandable, his comments are especially useful because they allow us to see that animals, especially large numbers of large animals convey the same sort of social, reputation-confirming power in the late Classical period that they did in Telemachos' time. In fact, Demosthenes has become so enraged because foreigners like Philip can use animals to elevate anyone to "true man" status. Birth, education, hard work, ancestral lands, none of these seemed to matter if the situation Demosthenes describes is accurate. The environmental factors that had limited the number of large-scale animal producers in Attika could not affect gifts like Philip's, at least not initially; in time, though, land and food would have to be found or the animals would have to be sacrificed or otherwise consumed in public displays.

And in Demosthenes' day, there were many public events at which animals could be used up, not only to show off but to build an impressive store of public goodwill. This must be the public envy and honor that Demosthenes complains about. Large animals such as cattle and horses allowed their owners to compete in exclusive, elite activities such as the pan-Hellenic horse and chariot races or large, privately-sponsored sacrificial dedications. These conspicuous displays were important aspects of 4th-century Athenian upper-class society and behavior, and as a result areas which the elite controlled very carefully. Philip's gifts, especially the horses, upset the local balance by allowing non-elites like Euthykrates to circumvent all local restrictions and exhibit traditional aristocratic forms of wealth, at least for a while. This sudden change in status seems to have especially scared Demosthenes because he

feared that more Athenians might follow Euthykrates'
example and seek similar social elevation in Philip's service.

Unfortunately for Demosthenes and his elite brethren,
other Athenians did seek to increase their animal wealth
and consequently their status, and not just from Philip of
Macedon. An Athenian cavalry officer named Meidias, for
example, while traveling on state business, attempted to
extort cattle and fence posts from Athenian allies.[46]
Although Meidias was a cavalry commander, and as such
probably already a wealthy man who did not need the
dramatic social elevation craved by Euthykrates, he could
still profit from the cattle. Cattle would allow Meidias the
necessary raw materials to make a dramatic public sacrifice
and thereby exhibit his gentlemanly resources. [47] Perhaps
even the fence posts would allow him to parcel off irrigated
land and establish a local herd. Whatever the case, Meidias
thought cattle worth the risk of prosecution and a stiff fine.
Like Telemachos of the *Odyssey*, Meidias needed animals
to demonstrate to others that he was, in fact, a gentleman
of wealth, worthy of their honor and envy.

In the actions of men like Meidias and Euthykrates, and
the resultant ire of Demosthenes, we see that Victor
Hanson was incorrect in his observation about the short-
lived importance of elite animal production. The fact that
Aristotle, near the end of the polis period, can argue for a
hierarchy of wealth that gives pride of place to horses,
followed by cattle, sheep, other animals, and finally
agriculture, suggests that Archaic and Classical gentlemen
placed extremely high value on their herds and flocks.

---

[46] Dem. 21.167.

[47] See chapter 5 for a detailed discussion of the impact expensive sacrifice
could have on an individuals public reputation, especially at the deme or local
level.

III

TENDING THE HERDS
ANIMAL MANAGEMENT STRATEGIES

One must be expert as to which animals are most profitable compared to each other, and also which are most profitable on what sorts of land, for different ones thrive on different lands.

Aristotle, *Politics* 1258b.15-17

And would not the same hold true in regard to sheep, if a man should suffer a financial loss because he does not know how to manage sheep, his sheep would not be a source of money for him either?

Xenophon, *Oeconomicus* 1.9

The lone cowboy, well known from television and the big screen, evokes a time when men roamed the wide-open ranges and cattle was king. Yet despite this strong, semi-mythical reputation, the reign of cattle in the American middle West was a relatively short lived phenomenon (a decade), having little or nothing to do with "lone cowboys." In fact, the cattle kingdoms of the 1870s were scarcely American productions at all, depending largely on the London financial futures' market and huge sums of money from Scottish, English, Dutch, and German investors for the purchase of land, livestock, fencing, and fodder. During the 1870s and 1880s these foreign investors were so feared that American state and territorial legislatures even passed laws circumscribing their influence, to little lasting effect. But inasmuch as foreign capital created the cattle kingdoms, it also facilitated their early demise: stockraising on the prairie was so dependent on the international economy, in the form of continued foreign investment and consumption, that it had no protection from the crushing depressions of the late 1870s and early 1880s. Once the influx of foreign capital slowed, the animals died in the fields through overgrazing, or at the stockyards, for want of a market.[1]

Such a dramatic example from frontier America serves to illustrate how connected animal production can be to specific political and social variables, how very much tied to a specific time and place. In much the same ways as King Cattle of the American West was a phenomenon of the 1870s financial futures' market, the elite production of cattle, goats, and pigs in democratic Athens was a result of the state-sponsored sacrifice market of the late fifth and fourth centuries BCE. The infrastructure that makes animal production systems possible—specialized markets, systems of landholding, investment by elites, state involvement—is a unique cultural construct, differing community by community. Consequently, largely because of these highly visible cultural characteristics, we should speak about systems of animal husbandry as discrete, historical and regional entities, differing from each other in significant and observable ways. Unfortunately, social and economic historians of Ancient Greece have treated ancient Greek animal husbandry as a uniform phenomenon, differing only in scale from producer to producer.[2] As

---

[1] R. G. Kennedy, *Rediscovering America* (Boston: Houghton Mifflin, 1990), 188f.

[2] S. Hodkinson, "Animal Husbandry in the Greek Polis," in C. R. Whittaker, ed., *Pastoral Economies of Ancient Greece and Rome* (Cambridge 1988), 35-74, and J. Skydsgaard, "Transhumance in Ancient Greece," in C. R. Whittaker, ed., *Pastoral Economies of Ancient Greece and Rome* (Cambridge: Cambridge University Press, 1988), 75-86. Even a recent work such as C. Chandezon, *L'élevage en Grèce (fin V^e-fin Ier s. a. C.)* (Bordeaux: Ausonius, 2003), which is sensitive to temporal and regional variation largely ignores the role of societal factors in shaping those variations.

discussed earlier, the two competing models of transhumance and argo-pastoralism have caused scholarship about animals in the ancient world to become fossilized and isolated from mainstream historical discussions. The following discussion attempts to highlight regional and temporal variation by showing through an analysis of four very different communities—Athens, Sparta, Thessaly, and Arkadia—that different individual elites devised unique ways, methods, and goals for keeping animals to meet local social, political, and economic agendas.

But before moving to the evidence, it is essential to confront some specialized terminology. Outside historical discussions, among the vast technical literature concerned with modern practical applications of animal husbandry, "animal management strategies" have come to describe the specific measures and techniques adopted by husbandmen in their effort to produce certain restricted end-products, such as milk, meat, or wool/hair.[3] Consequently, practical studies of modern animal husbandry often describe three broadly-defined strategies: (1) If the aim is to have animals for work and for the production of wool, hair and hides, then individuals tend to be kept until maturity; (2) if the aim is milk production—for which females are employed—then one sees a high reproductive rate, with most young males killed and females not kept beyond their reproductive years; (3) but, if the aim is meat production, young males are killed, sometimes after fattening, as juveniles or young adults, at the point when the most economical balance has been achieved between the weight of the animal and the feed consumed.[4] For the sake of clarity, this chapter will follow the trend already established in ancient historical scholarship and refer to

[3] See, for example, E. M. Ensminger, *The stockman's handbook* (7th ed.) (Danville, IL: Interstate Publishers, 1992), and *Beef cattle science* (7th ed.) (Danville, IL: Interstate Publishers, 1997).

[4] See G. Dahl and A. Hjort, *Having Herds. Pastoral Herd Growth and Household Economy* (Stockholm Studies in Social Anthropolgy 2) (Stockholm: University of Stockholm Press, 1976) for a detailed discussion of these strategies.

more general management typologies such as transhumance and agro-pastoralism as "animal management strategies," while reserving "animal production strategies" for the more specific and specialized techniques designed to produce wool, meat, or milk.

Since Sebastian Payne wrote a "reader's guide" to zoo-archaeology in Greece, scholars of ancient animal husbandry have been aware of the potential of animal bone analysis for illuminating both production and management strategies, but they have been equally aware of zoo-archaeology's specialized focus on prehistoric subsistence and taxonomy.[5] In 1994, a survey of "recent work in Greek zoo-archaeology" by David Reese confirmed that the discipline was still too specialized, too focused on the subsistence of prehistoric Greece at the expense of the more richly documented historical periods.[6] This emphasis on prehistory is unfortunate, since faunal reports, which document the age, species and general health of the animal at the point of death, could provide the most accurate picture available about how animals were kept and for what purpose. But there is some hope for the future. The emphasis on prehistory and subsistence has begun to shift, and most excavations of historical sites conducted in Greece now have a zoo-archaeological component. And in 1999 the first conference held in Greece on the subject of ancient Greek zoo-archaeology, acknowledged that more attention must be given to the historical periods. But the field of historical, non-subsistence zoo-archaeology is still in its infancy, and so for the foreseeable future, studies of

[5] S. Payne, "Zoo-archaeology in Greece: A Reader's Guide," in Nancy C. Wilkie and William D. E. Coulson, edd., *Contributions to Aegean Archaeology. Studies in Honor of William A. McDonald* (Minneapolis: University of Minnesota Press, 1985), 211-44. See M. H. Jameson, "Sacrifice and Animal Husbandry in Classical Greece," in C. R. Whittaker, ed., *Pastoral Economies of Ancient Greece and Rome* (Cambridge Philological Society, Suppl. Vol. 41), (Cambridge: Cambridge University Press, 1988), 87-119, for an assessment of the use of zooarchaeology for historical studies.

[6] D. Reese, "Recent Work in Greek Zooarchaeology," in P. Nick Kardulias, ed., *Beyond the Site. Regional Studies in the Aegean Area* (Lanham, MD: University Press of America, 1994), 191-223.

Archaic and Classical animal husbandry will remain confined to the literary and epigraphic sources.[7]

While modern animal production strategies are easily observable, and even those of the recent past, such as King Cattle, are well-documented, ancient strategies are rarely discussed or downright confusing in the literary sources. For example, Homer alludes to transhumance *and* fixed-based grazing, while Hesiod and Xenophon, whose *Works and Days* and *Oeconomicus* concern the rural lifestyle and might be expected to explain animal management strategies, devote only minimal attention to the technical aspects of agriculture or animal husbandry in general.[8] Witness Hesiod's overview of draft animals, oxen, and mules as being useful for arable farming.[9] Elsewhere, Hesiod makes equally brief mention of sheep, goats and cattle around the farm, freezing from the winter cold, and even lists certain "Days" that are related to animal husbandry: when the ram, boar, and bull should be castrated; when the sheep should be sheared; and when a heifer should be sacrificed.[10] Therefore, we see that domesticated animals do have a place on Hesiod's farm, but for whatever reason Hesiod simply does not wish to elaborate on how such animals are produced. Indeed,

---

[7] E. Kotjabopoulou, Y. Hamilakis, P. Halstead, C. Gamble, and V. Elefanti, *Zooarchaeology in Greece. Recent Advances. British School at Athens Studies,* 9 (London: The British School at Athens, 2003). See T. Howe "Review of *Zooarchaeology in Greece,*" *BMCR* 2004.03.07, http://ccat.sas.upenn.edu/bmcr/2004/2004-03-07.html.

[8] For mountain grazing, e.g. *Il.* 2.749; 5.315; 12.301; 16.352; 18.598; 21.448. For winter pasturage in the lowlands, e.g. *Il.* 17.549-50. Homer often pictures these flocks accompanied by their doughty shepherds *Il.* 13.492; 16.353; 11.106; *Od.* 15.386; 24.112;*Od.* 4.413. The poet even tells about sheepdogs (*Il.* 10.183) and painstakingly describes the many pens, shelters and corrals necessary for protecting the animals at night and during times of bad weather See *Il.* 18.588, for upland pastures and their equipment; Cf. *Il.* 8.131. For fixed grazing, e.g. *Il.* 2.775; 15.630-32; 20.221-222. In the marshy areas of the mainland, across from the Isle of Ithaka, Odysseus' herdsmen grazed his twelve herds of cattle, twelve flocks of sheep, and twelve droves of pigs, while only on rocky Ithaka itself did the hero keep his twelve herds of goats. *Od.* 14.99-104.

[9] *WD* 405, 436, 606.

[10] *WD* 786, 590.

Hesiod does not even describe the pens and corrals in which the draft animals were housed, nor the infrastructure for grazing sheep in the mountain pastures. This silence is especially puzzling since herding sheep on the slopes of Helikon played an important role in Hesiod's life, bringing him into contact with the Muses and starting him on his career as a poet.[11] Xenophon is even less helpful than Hesiod, yet animals must be present on Ischomachos' farm, for Sokrates earlier in the *Oeconomicus* (5.3) argues that the skill of animal production is closely linked with arable farming, and is necessary for the production of sacrificial victims.[12] It is perhaps due to this lack of detail that modern studies have overlooked the geographical diversity of animal production and instead come to view animal management in such general paradigms as agro-pastoralism and transhumance.

Only Aristotle's *Researches on Animals (Historia Animalium)*, which has been largely overlooked by historians of animal husbandry, offers a rich, nuanced, at times even technical discussion of animal production. For this reason, *Historia Animalium* is a good starting point not only for constructing a more balanced understanding of animal management but also for assessing the general state of Greek practical and theoretical knowledge about keeping animals during the Classical Period. But before discussing each species, it is necessary to draw attention to some general characteristics of *Historia Animalium*. First, Aristotle's main purpose is not to describe animal husbandry as such but rather to present a taxonomic classification and description of all living animals according to their similar characteristics.[13] Second, Aristotle has a gift

[11] *Theog.* 22. Cf. Paus. (9.31.2), who comments on the sheep and goat pastures of Helikon.

[12] But perhaps not to the degree Hodkinson, "Animal Husbandry in the Greek Polis," would have it. Ischomachos does not save the stubble after harvest so his animals can graze it. Instead he has it burned. Further, Ischomachos does not graze the fallow, instead having it plowed and dug by slaves (*Oec.* 16.10-15).

[13] Consequently, he divides domestic animals into two main categories: those living in herds, and those living in close connection to man. The dog and the pig are noteworthy of the latter, while horses, sheep, goats, and cattle are

for banal observation. In his discussion of cattle, for example, he advises that larger species require more extensive pastures.[14] Yet behind such obvious points lay a certain depth: the ancients understood the differences in land use in terms of carrying capacity and individual species' unique needs.[15] While most of Aristotle's statements, obvious or otherwise, are useful in some respect, a small number are painfully naïve and representative of folk tradition, such as his assertion that if sheep mate when the wind blows from the north, the offspring will be male, when the wind blows from the south, female.[16]

We begin our survey of the *Historia Animalium* with sheep, goats, and cattle, since these were essential to ancient social, religious, and economic life as sources of sacrifice, meat, hides, hair, and wool. These ruminants are grass eaters and Aristotle observes that they need a good deal of open range.[17] Sheep graze the pasture intensively, eating the grass and shrubs down to the ground, goats move around quickly and trim only the new shoots of the plants, and cattle must have rich, well-watered grazing in order to thrive.[18] Hence, goats require more land than

---

examples of the former. In fact, the herds of Epiros roam so far from the settlements of men that the bulls were not seen for three months at a time (*Hist. an.* 572b.20).

[14] *Hist. an.* 522b.20.

[15] Perhaps equally banal is the statement that all quadrupeds produce milk, but some produce more than is required for nourishing the offspring and this is used for cheese making (*Hist. an.* 522a.25-30). Yet even this observation tells us that the Greeks made cheese from the milk of certain domestic animals. In the same section, Aristotle also comments on the relative values of cheeses: The best is sheep's milk, next goats', then cows'. One *amphoreus* of goats' milk yields nineteen *obol*-cheeses while the same amount of cows' milk, thirty *obol*-cheeses. Peck in his Loeb translation suggests that *obol* here refers to the price of the cheese rather than the weight or shape.

[16] *Hist. an.* 574a. Modern stockmen are not immune to this sort of folklore. All have different techniques, often passed down through families, for ensuring twins and triplets among sheep, or males among cattle.

[17] *Hist. an.* 596a.10.

[18] Unlike herd animals, Aristotle advises that swine should live in close conjunction with humans and are most efficiently raised on human refuse (*Hist. an.* 596a.16-17), though they can be quickly fattened for market with

sheep but are less destructive to the plants, while cattle require irrigated land or supplemental feeding. And supplemental feeding seems to be a necessity for all three, at least for meat production; the philosopher preferred to eat animals which had been fattened in a controlled setting with the cereal residues, olive shoots, wild olive branches, vetch, grape pressings, and other types of vegetal byproducts.[19] While Aristotle lists many feeding strategies, the most important is salt additives. At the end of summer, probably to fatten for fall festivals, stockmen give their young charges salt every five days at the rate of one *medimnos* for one hundred animals. Although this technique does not result in actual meat production, it does dramatically increase water gain, which gives the impression of a fat, healthy animal; indeed, this may even be a means to maintain health through water retention over the long drive from feedlot to market during the hot temperatures of late Summer.[20]

Sheep, goats, and cattle are dependent animals, needing constant care, attendance, and a certain amount of protection from bad weather. According to Aristotle, they often leave their shelters in wintry weather and must be rounded up by their tenders. In order to make the round-up of sheep easier, shepherds regularly train a castrated ram while it is quite young to lead the others of the flock.[21] Goats, however, were not trained to follow a leader, because they are more individualistic in their grazing, and

---

barley, millet, figs, acorns, wild pears and cucumbers (*Hist. an.* 596a.18-595b.1).

[19] *Hist. an.* 595b-596a. But there was a risk that sheep can become too fat through overfeeding. At Leontini and on Sicily the shepherds do not turn out the sheep to the pastures until late in the evening, in an effort to reduce the amount they eat. (*Hist. an.* 520b. 1-3). Aristotle, it seems, could be sensitive to regional variations in animal production but he mentions such differences rarely, and then only as extreme illustrations.

[20] *Hist. an.* 596a.10-24. Many ranchers in the Western USA feed their animals salt, so they will take up extra water and thus fare better during the long, hot ride from pasture, or feedlot, to market.

[21] *Hist. an.* 573b.25-27.

cattle tended to stay in herds with no designated leader.[22]
The shepherds also trained their sheep and goats to become
accustomed to sudden noises, so they would not become
unduly frightened by a thunderstorm and consequently
miscarry if pregnant.[23] Such was the devotion of shepherds
to flock that at night they even slept in the shelters with
their animals to protect them from cold and predators.[24]

Unlike animals used for human consumption, horses,
mules, and asses are not fattened. Consequently, Aristotle
recommends that they graze in large herds, out on open
pasture and with only the young taken for training and
sale.[25] In order to ensure a good strong animal, however,
the horseman should plant and maintain alfalfa, since it
makes horses sleek and strong.[26] For best results, equines
are kept in large herds, and special horse-trainers are
employed to maintain the herds and separate out young
animals, when their time comes for training. Unlike
shepherds or cowherds, horse-trainers do not choose a
leader for the herd, nor manage the herd too much, yet the
trainers do supervise their charges, making sure their herds
are healthy and manageable (40 individuals or less under a
dominant stallion).[27]

While Aristotle provides more technical information
than Hesiod or Xenophon, and offers a useful point of
departure for a full view of animal production, in terms of
general practice and ideal goals, he (like many modern
scholars) is only concerned with production of animals in a
taxonomic sense. In order to grasp the methods and scope

[22] *Hist. an.* 574a. 11-12; 611a.7-9.

[23] *Hist. an.* 611a. 4-5. This is a real fear among stockmen. From my own
experiences herding sheep, I can recall losing lambs on at least three separate
occasions where loud noises such as thunder or gunshot caused pregnant ewes
to miscarry, resulting in the death of their lambs.

[24] *Hist. an.* 610b. 30f. Woolly sheep do not winter as well as the broad-
tailed variety (*Hist. an.* 596a.25-596b.8).

[25] *Hist. an.* 611a.10-15.

[26] *Hist. an.* 596b.23-29.

[27] *Hist. an.* 572b.10; 577a.15-18. The ass is raised much like the horse, but
lives on less feed. Aristotle also includes much technical data about the proper
crossbreeding of horses and asses to produce draft mules.

of animal management strategies, as practiced during the Classical period, we need to explore the interaction of variables such as landscape, fodder, market demand, and socio-political systems. Consequently, we turn to the two best-studied *poleis*, Sparta and Athens, and contrast them with the *ethne* of Thessaly and Arkadia.

Athens

> Now the Athenians of old fought the wolves, since their country was better for grazing than farming.

Plutarch Solon 23.4

As Plutarch says, the Athenians were given a land better for grazing than farming, and Athenian gentlemen could make fortunes in both.[28] In his typology of wealthy Athenians, Xenophon ranks sheep ranchers before wine, oil, and grain farmers.[29] The value of these Athenian sheep derived from their high quality wool; Polykrates, tyrant of Samos, was so impressed with Attic wool that he wanted to import Athenian animals for crossbreeding with Milesian sheep, in an attempt to improve his own herds and thereby compete with, or perhaps surpass, the Athenians.[30] Indeed, by the Roman period, all knew the quality of Attic wool: "What other wool is softer than Attic?" is recorded by Athenaios as a most foolish question.[31] But in addition to wool, the Athenians also exploited their sheep for cheese, and Athens was famous for its fresh cheese market, frequented during the fifth century BCE by men from outlying areas as far away as Plataia.[32]

---

[28] See Bacchyl. 18.9.

[29] Xen. *Poroi* 5.3. Lysikles "the sheep dealer," a contemporary of Perikles may have been one of these (Ar. *Eq.* 132, cf. 739).

[30] Athen. 12.540d. Wool is listed as a particular product of Athens by Antiphanes, a fourth century BCE playwright (Athen. 2.43c).

[31] Athen. 5.219a.

[32] Lys. 23.6. It is probable that men in rural areas like Plataia raised the sheep that produced the cheese sold at this market. Such a practice would certainly account for Plataian interest in the Athenian cheese market. Cf. Ar. *Eq.* 479-80, where Athenians are selling cheese as far afield as Boiotia.

The Attic orators provide some clues concerning the practical management of these sheep. [Demosthenes] 47 describes a herd of 50 fine-fleeced animals, supervised by both a herdsman and assistant, which were stolen by rustlers while pastured out, away from the cultivated farmland.[33] That the speaker identifies them as wool breeds, that such a small flock has two shepherds, and the fact that they were stolen, suggests these were valuable animals. But such care is not unique—wool sheep in Attica and the Megarid usually wore leather coverings, or "jackets," to protect their wool from dirt and the elements, and thus ensure a better price at market.[34]

The orator Isaios shows the other main concern of Athenian husbandry, goat production. He describes the affairs of Euktemon, who owned a herd of goats, together with their herdsman, which was valued at 1,300 drachmas, a not inconsiderable portion of an estate worth 3 talents.[35] While the speaker of Isaios 6 does not tell us the number of goats, or even the reasons for which they were kept, he does suggest that they provided their owner a ready income with which to engage in important public duties such as liturgies. It is likely that Euktemon was selling off young animals for the sacrificial/meat market, and thus engaged primarily in meat and hide (perhaps also cheese)

[33] It seems clear that these sheep were grazing under supervision, perhaps on nearby the hillsides, just as Aristotle recommends in the *Historia Animalium*, and were only loosely connected to the farm at the time when they were seized, because as [Demosthenes] carefully observes, the animals were taken *before* the rustlers could trespass on the plaintiff's land, as Isager and Skydsgaard *Ancient Greek Agriculture. And Introduction* (London: Routledge, 1992), 102, argue. Hodkinson, "Animal Husbandry in the Greek Polis," had argued the opposite, that the owner of the 50 sheep was engaging in mixed, agro-pastoralist strategies (i.e. housing his animal in stalls at night and supplementing their grazing with agricultural byproducts), since the sheep were grazing in close proximity to the owner's farm when stolen. These are not mutually exclusive. The sheep are disconnected at the time of the crime from the main agricultural base, but whether this is usual or part of some complex strategy is unknowable, for the speaker of [Dem.] 47 is not really concerned with how the sheep were raised. Instead, he concentrates on his main theme: the fact that the sheep and their shepherds had been taken by force, against their owner's will by his creditors.

[34] Diog. Laert. 6.41.

[35] Is. 6.33.

production.[36] The Athenian demand for sacrificial animals
was so great in the 4th century that it effectively created its
own market: 6,528 oxen and 15,186 sheep/goats were the
minimum numbers of animals required yearly for both the
deme sacrifices and the *epithetoi heortai*, the large state-
sponsored sacrifices of the late-fifth and fourth centuries.[37]
And Euktemon was not alone in recognizing the potential
of this market. A miller named Nausikydes invested the
money earned from his mill in a large herd of pigs and
cattle. The income derived from selling the animals was so
great that Nausikydes was able to support his family and
even undertake expensive liturgies.[38] The wealth-
generation potential of animal production, and social power
of the liturgies it could buy, seems to be the crux of
Demosthenes' complaint when Philip of Macedon rewarded
several non-elite Athenian ambassadors with a number of
sheep, cattle, goats, and horses.[39] Because of their animal
wealth, and the opportunities it gave them, these men were
able to become leaders of the polis, honored and even
envied.[40] During the Classical period, state sacrificial
demand created an artificial situation in which large profits
could be made selling sacrificial victims, profits large

---

[36] S. Hodkinson, "Imperialist Democracy and Market-Oriented Pastoral
Production in Classical Athens," *Anthropolozoologica* 16 (1992): 53-61.
According to Isaios, these goats were just one of Euktemon's many sources of
income; he also rented out both city and farm property.

[37] V. Rosivach, *The System of Public Sacrifice in Fourth-Century Athens*
(Atlanta: Scholars Press, 1994), 78, n. 27. In the fourth century, largely because
of the sacrificial market, Attika experienced the highest level of intensive
farming coupled with animal production. This was abandoned abruptly at the
end of the century and less intensive methods, such as transhumance took its
place. See H. Lohmann, "Agriculture and Country Life in Classical Attika," in
B. Wells, ed., *Agriculture in Ancient Greece* (Stockholm: Paul Astrom, 1992),
29-57, for discussion.

[38] Xen. *Mem.* 2.7.6.

[39] Dem. 19.265. Isaios 11.40-43 probably describes the size of a typical elite
holding: 60 sheep, 100 goats, and a cavalry horse.

[40] Plato, in the *Laws* (743d), like Demosthenes, criticises those who made
great profits by fattening castrated animals for the sacrificial meat market. See
J. K. Davies, *Wealth and the Power of Wealth*, (Salem, NH: Arno, 1981), for a
discussion of the social power liturgies can buy.

enough to entice even wealthy citizens like Nausikydes to invest in animal production.

While the forensic sources do not always specify how these animals were raised, we can infer a great deal from how the animals were themselves described. The sheep and goats in the above examples are all treated separately, as herds, usually together with their shepherds, rather than as part of a working farm.[41] Recent archaeological evidence from the Attic countryside seems to support such conclusions about semi-mobile herds of sheep and goats, grazing marginal pastures at some distance from the main agricultural base. In such cases, the remote buildings and tower constructions scattered across the Attic countryside served as herding stations, complete will corral compounds, to provide nighttime protection.[42] The Athenians would have sent their flocks into the hills for extended periods of time in order to exploit the grazing these areas offered. The well-pastured mountains near the borders of Boiotia and especially the Megarid are thick with such pastoralist sites.[43]

[41] The sheep and goats mentioned in the Attic Stelai were also listed in this fashion (*IG* I^3 426.58f.). The 67 goats and 84 sheep are registered with their young, separate from other property.

[42] The Cyclops in Euripides' *Cyclops* grazed his animals on the slopes by day but penned them each night. This seems to have been the practice in Southern Attika, with animals kept close to farm complexes. In the northwest, though, herding stations tended to be remote, removed from cultivated areas and devoted primarily to animal production. For a discussion of the debate surrounding the use of towers in southern Attika see R. Osborne, "'Is it a Farm?' The Definition of Agricultural Sites and Settlements in Ancient Greece," in B. Wells, ed., *Agriculture in Ancient Greece* (Stockholm: Paul Astrom, 1992), 21-28, and H. Lohmann, Agriculture and Country Life in Classical Attika." For northwestern Attika and the Megarid see H. Lohmann, "Antike Hirten in Westkleinasien und der Megaris: zur Archäologie der mediterranen Weidewirtschaft," in Walter Eder and Karl-Joachim Hölkeskamp, eds.,*Volk und Verfassung im vorhellenistichen Griechenland* (Stuttgart, Franz Steiner Verlag, 1997), 63-88.

[43] Thucydides (5.42) observes that the border between Attika and Boiotia at Panakton was a recognized grazing ground from ancient times. Lohmann "Antike Hirten," 75, adds that the ancient Megarid may have been even more "pastoral" than Attika, and longer distance movements, which exploited seasonal grazing, more the norm in that area. Clearly, Megara was an important sheep producing region. In antiquity the area boasted an important sanctuary to Demeter the Sheepbringer and was known for its quality woolens.

Nonetheless, the degree of mobility would have depended on the resources the individual owner could control and on the number of animals he owned. The property boundary stones and rock-cut inscriptions found in the Attic countryside suggest that many Athenians protected grazing in otherwise predominantly agricultural regions. These *horoi* boundary stones are usually found along the steep ridges which separate one farm from another. As such they are in plain sight to those who approach the crest of the ridge, and as G. Stanton put it, they seem to be saying "Don't bring your sheep or goats over here."[44] Perhaps if an owner could control the hillsides around his farm, he might not need to lease pasture or move his animals farther from his agricultural base.[45]

Controlling land is essential to large-scale, elite animal production, and at Athens, as elsewhere in Greece, socio-political conventions affected the amount of land a man could control. At least from the time of Solon, Athenian landholding was a complicated affair, with all citizens having potential access to land, and most owning at least one plot, albeit small.[46] As a result, because of inheritance and other social factors, the Attic landscape became subdivided over time to a degree seldom seen elsewhere in the Greek world, consisting of an intricate patchwork of

---

Ar. *Acharn.* 519; Xen. *Mem.* 2.7.6; see R. P. Legon, *Megara* (Ithaca, NY: Cornell University Press, 1981), and E. Mantzoulinou-Richards, "Demeter Malophoros: The Divine Sheep-Bringer," *AncW* 13 (1986): 15-22.

[44] G. Stanton, "Some Attic Inscriptions," *ABSA* 79 (1984): 289-306, and "Some Attic Inscriptions," *ABSA* 92 (1997): 178-204, suggests that these *horoi* protected deme land. He observes that two in particular seem to mark the boundary between Coastal Lamptrai and Upper Lamptrai. As a result, these particular *horoi* probably defined areas of public grazing leased out by the demes, not lands held by private citizens.

[45] For a general discussion of lease land see D. Lewis, "The Athenian *Rationes Centesimarum*," in M. I. Finley, ed., *Problèmes de la terre en Grèce* (Paris: Mouton & Co., 1973), 187-212, and R. Osborne, Demos: *The Discovery of Classical Attika* (Cambridge: Cambridge University Press, 1985), 56-59.

[46] T. W. Gallant, *Risk and Survival in Ancient Greece. Reconstructing the Rural Domestic Economy* (Stanford: Stanford University Press, 1991); V. D. Hanson, *The Other Greeks* (Berkeley and Los Angeles: University of California Press, 1998).

small, individually owned parcels.[47] This hodgepodge character is best illustrated by the property liens discussed by Finley in his classic study of Athenian credit and landholding and the Attic Stelai, or lists of property confiscated from those who mutilated the herms and profaned the Mysteries with Alkibiades in 415 BCE.[48] Such a fragmented system of land tenure and land use certainly had an effect on the number of animals that Athenians could keep in any one place. Consequently, one sees individual flocks not much larger than 50 sheep, about the amount that one shepherd (or a shepherd and his assistant, if they were especially valued) could reasonably herd. In practice, wealthy Athenians may well have kept many such herds of sheep on their scattered properties, but they could not have grazed larger herds on individual holdings. Only in the border areas, or on the uninhabited slopes of the larger mountains such as Pentelikon, Hymettos, or Parnes, where Lohmann identified the large herding stations, might several herds be kept.[49]

Inasmuch as Athenian practices of landholding profoundly affected the ways in which flocks were managed in Attika, climate and geography must also have played a role, if only in limiting the types of animals the Athenians could raise. Since Attika is one of the driest regions of mainland Greece, with very little wetland pasture, large animals such as cattle and horses, which require abundant fodder and water, would not prosper, and certainly would not survive in the big herds suggested by Aristotle.[50] As a result, as Plutarch observes (*Solon* 23.4),

[47] R. Osborne, *The Discovery of Classical Attika*, 47-63. B. Wells, ed., *Agriculture in Ancient Greece* (Stockholm: Paul Astrom, 1992). N. F. Jones, *Rural Athens Under the Democracy* (Philadelphia: University of Pennsylvania Press, 2004).

[48] *IG* I³ 426. M. I. Finley, *Studies in Land and Credit in Ancient Athens 500-200 B. C.: the horos inscriptions* (New Brunswick, NJ: Rutgers University Press, 1952).

[49] Perhaps it is significant that Euktemon, the owner of 1,300 drachmas worth of goats, had property on the slopes of Pentelikon (Is. 6.33).

[50] Osborne, *Classical Landscape with Figures* (London: George Philip, 1987), and P.D.A. Garnsey, *Famine and the Food Supply in the Graeco-Roman World* (Cambridge: Cambridge University Press, 1988), 89-106. It is telling

Attika is goat and sheep country, with the larger animals kept only in small numbers. The Attic Stelai reflect this reality: the estate of Panaitios, for example, contained only 2 draft oxen, 2 unspecified oxen, 4 cows with an unknown number of calves, 67 goats and 84 sheep, together with an unregistered number of offspring.[51] This lack of large animals in Attika may explain why Solon forbade the sacrifice of oxen at funeral feasts, and why in the Athenian sacrificial calendar sheep were regularly substituted in the place of oxen.[52]

As for horses, although a cavalryman and horse-racer like Xenophon advises against home production, Athenians did retain some horses, though never in large herds, for the Athenian cavalry on active duty were required to maintain their horses on their own land (about 700-1200 horses in total, excepting replacements).[53] Stratokles, the man who profited from his niece's dowry, kept a cavalry horse.[54] He could afford to do this because the state supplemented cavalry fodder expenses by providing a cash advance to each horseman for the purchase of grain.[55] This grain

---

that Thucydides (2.14.1) lists only draft-animals and sheep when describing the livestock the Athenians evacuated to Euboia during the Peloponnesian War. See V. D. Hanson, *Warfare and Agriculture in Ancient Greece* (Berkeley and Los Angeles: University of California Press, 1983), for discussion of the wartime evacuation and its effect on agricultural production. The lease inscriptions from demes and rural sanctuaries seem to support a shortage of animals in the Classical period. *IG* II² 2493, for example, expressly prohibits the taking of manure from the lease land, suggesting a general shortage of manure at the time. See Osborne, *The Discovery of Classical Attika*, for further discussion.

[51] *IG* I³ 426.58ff. The estate of Theophon, discussed above, contained only one horse.

[52] Plut. *Sol.* 21. See S. Dow, "The Greater Demarkhia of Erkhia," *BCH* 89 (1965): 180-213, for a discussion of the sacrificial calendar at Athens. Rosivach, *Sacrifice*, suggests that these substitutions were even more common than Dow suggests.

[53] *Oec.* 3.10.

[54] Is. 11.40-43.

[55] I. G. Spence, *The Cavalry of Classical Greece* (Oxford: Oxford University Press, 1993), Appendix 4, argues that each cavalryman was given what his particular horse might cost, up to 1,200 drachmas. G. R. Bugh, *The*

supplement substantially reduced the amount of grazing land an owner might need, and thus allowed Athenians to keep horses in their pasture-poor environment. Here again we see social institutions creating an opportunity for animal production where environment would have discouraged it.

In the end, we can conclude that the Athenians were primarily sheep and goat producers. Their production strategies, shaped by local systems of landholding, environmental limitation, the sacrificial market and concomitant elite need for wealth to fulfill their competitive liturgical obligations, seem to have been particularly varied. Perhaps the oath of the Furies, from Aischylos' *Eumenides* (938-46) best summarizes the agricultural cares of the Athenians:

> May leaf-destroying ruin not blow
> (I speak graciously)
> its blasts of heat, stealing buds from plants,
> lest they pass the border in these places;
> may no deadly plague draw near to kill the crops;
> may Pan at the appointed time
> nurture the thriving flocks with twin offspring.

Sparta

> You have only to look at the wealth of the Spartans and you will see that wealth here is far inferior to the wealth there. Think of all the land they have both in their own country and in Messenia, not one of our [Athenian] estates could even compete with theirs in extent and excellence, nor in ownership of slaves and especially of those from the helot class, nor yet of horses, nor of all the flocks that graze in Messenia. Plato, *Alkibiades* 1.122d-e

Spartan animal management strategies differed greatly from those of the Athenians mostly because of Spartan

---

Horsemen of Athens (Princeton: Princeton University Press, 1988), 53-58; 66-70, suggests that each cavalryman received a set fee of 1,200 drachmas.

social institutions. At Sparta a narrowly-defined elite controlled all property and resources.[56] In addition, the well-watered river valleys of Lakonia and Messenia, in contrast to the dry plains of Attika, enabled the Spartans to keep horses and cattle on a scale simply not possible at Athens.[57] Euripides (*apud* Strabo 8.5.6), for example, speaks eloquently about the resources and gentle climate of the region:

> Watered by countless streams,
> furnished with good pasture for both cattle and sheep,
> being neither very wintry in the blasts of wind,
> nor yet made too hot by the chariots of Helios.

Consequently, because of the unique social, political, and environmental characteristics of Sparta, the Spartan landscape was dominated by large estates and large herds rather than a patchwork of small units.[58] The Spartans also had a compelling social market for meat and other pastoral products. The public mess, to which every Spartiate must contribute or lose his citizenship seems to have served much the same role as obligatory liturgies in Athens, providing the wealthier Spartiates with a competitive forum in which to demonstrate the calibre of their wealth, through gifts of meat and cheese.[59]

The Spartans also competed in horse production and had demonstrated a passion and no small skill for chariot-racing; between the years 448-420 BCE a Spartan won the four-horse race in seven out of eight Olympiads. Aside

---

[56] See Arist. *Pol.* 1270a15-b6. For a discussion of the Spartan system of land holding see S. Hodkinson, "Land tenure and inheritance in classical Sparta," *CQ* 36 (1986) 378-406.

[57] Homer praises the green meadows of Lakonia and Messenia: *Il* 3.74; 4.530; *Od.* 4.602-604.

[58] Xenophon's comments (*An.* 5.6.4) about his estate at Skillous in Elis, which he dedicated to Artemis, suggests that this sort of large estate, in which agricultural land as well as pasture was individually owned, was not uncommon in the southern and western Peloponnese. See Xen. *Hell.* 3.2.26 for the vast numbers of sheep and cattle in Elis. Cf. Theoc. 25; *Il.* 23.202-203; 11.671-680; 9.707-723, 739-747.

[59] Athen. 4.139c, 140c-e, 141e. See S. Hodkinson, "Social order and the conflict of values in classical Sparta," *Chiron* 13 (1983): 239-81.

from raising animals for competition, Xenophon tells us that each Spartan was expected to provide his own horses for the cavalry forces, though in practice the wealthier Spartiates often loaned out horses to their less wealthy peers as a way of creating social obligations.[60] Indeed, the Spartans were so successful at, and famous for, their horse production that in the Hellenistic period they even exported animals to the Ptolemaic Kingdom of Egypt.[61] While the literary sources do not provide a clear picture of how these horses were raised, it is probable that the Spartans kept herds much as Aristotle described, on the well-watered pastures of their estates, complete with helot grooms and trainers.[62]

Apart from their horses, the Spartiates were also famous for their cattle.[63] In fact, King Agesilaos II took such pride in his herds that he presented each new member of the *Gerousia* with an unblemished ox when they entered office.[64] The Spartans also seem to have possessed smaller animals in abundance, raised primarily in large herds. The Platonic *Alkibiades* in particular contrasts the numerous flocks and the large estates of Messenia with the smaller scale of Athenian herds and lands. Yet as with Athens, the degree of integration between arable farming and animal husbandry remains elusive, though the well-described estate of Xenophon at Skillous in Elis may offer a useful context against which to evaluate the Spartan evidence, since Skillous seems to echo the rich estates described by the Platonic *Alkibiades*. Xenophon tells us that the grazing resources of Skillous were so abundant that the visitors to the local festival of Ephesian Artemis (whose shrine and festival Xenophon established and continued to support) could even pasture their sacrificial livestock and beasts of burden while attending the ceremonies. The meadows and

[60] Xen. *Hell.* 4.4.10-11; 6.4.11; *Lak. Pol* 6.3; Arist. *Pol.* 1263a35-6.

[61] Polyb. 5.37.

[62] Agesilaos II, for example, was said to have stocked his estate with many horses (Xen. *Ages.* 9.1).

[63] Strabo 8.5.6.

[64] Plut. *Ages.* 4.3.

tree-clad hillsides were excellent for raising pigs, goats, cattle, and horses.[65] The overall impression given by Xenophon is that all types of animals were raised at Skillous in large quantities, each in its own distinct enclave, seemingly independent from arable farming.[66]

Although the Spartan estates may well have possessed all the grazing and fodder resources necessary for on-site animal production, some form of seasonal movements were employed for sheep and goats. Indeed, a dispute over mountain grazing seems to have precipitated the First Messenian War, the war in which Sparta began to subjugate Messenia. The accounts agree that the hostilities began when the Spartans encroached on some borderland near Mount Taygetos. The exact nature of this encroachment is unclear, but all reports agree that the Spartans seized the borderland for their own use and set up a shrine to Artemis of the Wetlands. The fact that the Spartans dedicated the land to Artemis of the Wetlands suggests that this area contained springs and pasturage and therefore, the dispute centered around access to grazing.[67]

In the end, the abundant, well-watered plains of Messenia and Lakonia, with their large amounts of farming and grazing land, worked by the servile helots, witnessed a greater separation between agriculture and animal

[65] *An.* 5.3.11-12.

[66] Any sort of integration between pastoral and agricultural spheres of production would be minimal, probably even less than H. A. Forbes, "The identification of pastoralist sites within the context of estate-based agriculture in ancient Greece: beyond the Transhumance versus Agro-pastoralism debate," *ABSA* 90 (1995): 325-338, postulated for the Argolid, because Xenophon (and his Spartan friends) was not under the pressure to develop alternate sources of fodder as the Athenians and elite producers from drier regions such as Argos. Skillous possessed abundant, year-round grazing. Any supplement to the natural graze would be on an *ad hoc* basis, when there were agricultural byproducts near at hand.

[67] Tac. *Ann.* 4.43; Paus. 3.7.4; 4.31.3; 4.4.2. Before they had the rich, well-watered Messenian Plain, the Spartans may have needed the heights for their sheep. There is no indication, however, that Spartan exploitation of the upland areas diminished after the conquest of Messenia. Fifth- and fourth-century Spartan incursions into areas such as Thyrea suggest that acquisition of grazing land still was a major concern. Hdt. 1.82; cf. *Anth. Pal.* 7.244; 7.431-2; Thuc. 2.27.2; Eur. *Elec.* 413.

husbandry than existed anywhere in Attika. The Spartans had land and labor to spare, with a relatively small landowning class competing for available resources. Indeed, the production of animals for food was a social necessity at Sparta since all Spartiates had to contribute to the mess, with the wealthier citizens competing to see who might donate the most meat and cheese and thereby gain the most respect.

A Lowland Ethnos: Thessaly

> To the Thessalians it is seemly for a man to select horses and mules from a herd himself and train them, and also to take one of the cattle and slaughter, skin, and cut it up himself, but in Sicily these tasks are disgraceful and the work of slaves. *Dissoi Logoi* 2.11[68]

The Thessalian elite, like their counterparts in Sparta, controlled large tracts of well-watered land, worked by a dependent population. The one thousand cattle and ten thousand sheep, goats, and swine that Jason, the tyrant of Pherai collected from his subjects in 370 BCE to offer at a single Pythian festival suggest that Thessalians raised animals on a grand scale. And Jason's sacrifice would have only represented the surplus each elite producer could give, since the individual owners would need to retain a viable number of animals in order to maintain the health and productivity of their herds.[69] When speaking of the abundance of Thessaly, Theokritos recalls the immense herds of the Homeric epics, and even evokes Homeric language when he praises the attendants and cattle of the Skopadai clan and numerous sheep raised by the Kreonidai clan.[70] These large, well-tended Thessalian herds seem to be the inspiration behind many discussions in Aristotle's *Historia Animalium*.

---

[68] Translation R. K. Sprague, *The Older Sophists* (Columbia, SC: University of Missouri Press, 1972), 282.

[69] So Xen. *Hell.* 6.4.29 asserts.

[70] 16. 34f.

And as with Sparta and Athens, social incentives underlay the Thessalian animal production. The *Dissoi Logoi* suggest that Thessalian elite considered it a point of honor to be personally involved with animal production. Inasmuch as the social value the Thessalian elite placed on animal wealth determined the numbers of animals required, the environment of Thessaly shaped the methods of that production. The marshy areas of the Thessalian and Malian plains offered areas of superb grazing, with many regions fit only for animal production, since they lacked the drainage necessary to support grain crops.[71] Thus, to a greater degree than Greeks elsewhere, the Thessalians could specialize in animal production. It is, then, no surprise that the inhabitants of Thessaly were famous for their large herds of pasture-intensive cattle and horses.[72] Indeed, Thessalian stud farms bred Alexander the Great's famous horse Bukephalos and supplied the chariot horses which carried Orestes to victory at Delphi in Sophokles' *Elektra*.[73]

The Thessalians must have been in Aristotle's mind when he wrote of large herds of horses, cattle, and sheep, all supervised by grooms, trainers, shepherds and cowherds. Yet these animals need not have always been kept off arable land. As Aristotle observed, the ancient Greeks often fattened their animals on agricultural refuse, and the Thessalians in particular were adept at integrating pastoralism into their other agricultural strategies, employing the practice of tillering, or winter grazing of

---

[71] Aristophanes in the *Frogs* (1384) speaks about the many cattle grazing along the Spercheos river in Malis. Conversely, parts of the plain may have been too dry to graze animals, requiring some sort of transhumance. Livy (42.57) tells of regions particularly affected by seasonal drought.

[72] Many authors speak of "well-flocked" Thessaly: Bacchyl. 14b.6; *Il.* 2.696; 9.446; *Od.* 11.257; Strabo 6.5.18. Homer (*Il.* 18.573-576) praises the oxen, the horses and the harvest of Thessaly. For sheep and cattle, see Theoc. 16.36-9. H. D. Westlake, *Thessaly in the Fourth Century B.C.*, (Oxford: Oxford University Press, 1935), 4f., observes that the coinage of Larisa in particular bears cattle and sheep motifs. For horses see Soph. *Elec.* 703-6; Eur. *Andr.* 1229; Plato *Leg.* 625d; Theoc. 18.30. For the famous Thessalian cavalry see Hdt. 7.196; Isoc. 15.298.

[73] Arr. *An.* 5.19.4-6; Soph. *Elek.* 703-4.

grain crops, to slow the maturation of the grains and also substantially increased their yield.[74] Consequently, small animals such as sheep and goats grazed the lowland grain fields in winter, in addition to the unfarmed areas, and then moved to the abundant summer pastures in the surrounding high mountains and river basins when the crops began to mature. Cattle and horses, however, while they might have grazed the grainfields in winter, were in all likelihood never far from the lush pastures of cultivated alfalfa or the year-round wetlands.

The Upland Ethne: Arkadia and Central Greece

And he [Pan] came to Arkadia, land of many springs and mother of flocks.

Homeric Hymn to Pan, 30

As with lowland Thessaly, the mountain dwellers of Phokis, Lokris, and Arkadia also possessed the necessary geographical/ecological conditions to develop more specialized forms of pastoralism, less connected to arable agriculture than those of the Athenians or even the Spartans. Because of the general shortage of quality arable land outside of the lowland river valleys, the shorter growing season, and the difficulty of raising the primitive wheats and barleys at the higher elevations,[75] the upland communities developed pastoral production. The highland meadows offered abundant summer pasturage for sheep

---

[74] Theophr. *Hist. pl.* 8.7.4. Thessaly was famous, too, for its large yields of grain (Xen. *Hell.* 5.4.56-7). Tillering is also known during the Roman era and was greatly praised. K. D. White, "Wheat Farming in Roman Times," *Antiquity* 37 (1963): 209, and *Roman Farming* (Ithaca, NY: Cornell University Press, 1970), 134.

[75] The problem is one of growing season. The primitive grains required a longer season than the hybrids of today. As P. Garnsey, "Mountain Economies in Southern Europe," in C. R. Whittaker, ed., *Pastoral Economies of Ancient Greece and Rome* (Cambridge: Cambridge University Press, 1988), 196-209, points out the growing season falls from 170 days at 1,000m (Pindos in the Tetrapolis of Doris) to only 95 day at 2,000m (the highland meadows of Parnassos, Kiona and Vardousi); even modern hybrids have trouble growing above 1,800m. See R. Sallares, *The Ecology of the Ancient Greek World* (Ithaca, NY: Cornell University Press, 1991), 309.

and goats, free from the farmer's plough. In fact, the ancient sources are explicit about sheep grazing the high slopes of Parnassos from the earliest times,[76] and a recent archaeological survey of modern and ancient settlement in the eparchy of Doris, conducted by P. K. Doorn has suggested that communities with a majority of territory above 1,200m were dependent primarily upon stockbreeding, while those below 1,200m, primarily upon arable and arborial agriculture.[77] These highlanders exchanged wool, cheese, and even the animals themselves, in return for the agricultural produce of their lower neighbors. In fact, Doorn concludes that both lowland and highland communities depended on these periodic exchanges of resources.[78]

Because of the central importance of animal production among the upland communities, mountain pasturage was a constant source of contention. Witness the early fourth century BCE dispute between the Lokrian and Phokian communities around Parnassos. The Oxyrhynkhos Historian observes that these two groups were in a state of continuous raiding and petty warfare, forever stealing the sheep, and the grass, of their neighbors.[79] Elsewhere on the Parnassos massif, in the upland pastures near modern Arakhova, the Ambryssians and Phlygonians carefully walled off their pastures in a fashion which recalls the

[76] Eur. *Andr.* 1 100-1 has sheep pasturing on the highlands of Parnassos, as does *Hymn. Hom. Ap.* 303f. and *Il.* 9.406. The quality of this pasturage is disputed. See J. Skydsgaard, "Transhumance in Ancient Greece," in C. R. Whittaker, ed., *Pastoral Economies of Ancient Greece and Rome* (Cambridge Philological Society, Suppl. Vol. 41) (Cambridge: Cambridge University Press, 1988), 75-86 and S. Hodkinson, "Animal Husbandry in the Greek Polis," in C. R. Whittaker, ed., *Pastoral Economies of Ancient Greece and Rome.* (Cambridge Philological Society, Suppl. Vol. 41) (Cambridge: Cambridge University Press, 1988), 35-74, for differing views.

[77] L. S. Bommlejé and P. K. Doorn, eds., *Stroúza Region Project (1981-1993): an historical-topographical fieldwork* (Utrecht, 1984) 14-15, 28f.

[78] Pausanias observes how the fair stimulated by a festival of Isis in Southern Phokis provided all the surrounding pastoralists a ready, seasonal market in which to sell their animals as sacrificial victims (Paus. 10.32.15).

[79] In one instance these raids escalated into a pan-Hellenic war (the Korinthian War of 385) when each side called for support from allied states. *Hell. Oxy.* 18.3-4.

horoi of Attika, or, better yet, the barbed-wire of frontier America.[80] Yet not all communities fought over pasture. The towns of Myania and Hypnia agreed, among many other things, to set aside land for common use in order to allow shepherds from each community pasturage while in transit from upland and lowland ranges.[81] The pasture, however, was common *only* among the citizens of the two participating communities, and guards, paid by a special "pasture tax," patrolled the boundaries and evicted outsiders.

The centrality of animal husbandry to the peoples of the uplands is even better illustrated in western Arkadia. To a greater degree than anywhere else in Greece, the highlands of Arkadia are unpopulated, with the only permanent structures religious sanctuaries. M. Jost has observed that these isolated sanctuaries served as community centers, as meeting areas from which the Arkadians exploited the pastoral landscape, giving the dispersed, mobile, Arkadians focal points for their pastoral lifestyle. Such a sense of community would be quite different from that of more settled, urban folk such as the Athenians. In fact, even when the western Arkadians did move into large urban centers like Megalopolis, they did not abandon their rural shrines but instead created new festivals and urban sanctuaries twinned with their rural predecessors, which continued to stress the close connections between Arkadian life and the pastoral countryside.[82]

One of the primary gods worshipped in these rural sanctuaries was Pan, patron god of shepherds. Consequently, it is no surprise that in antiquity Arkadia was famous for its sheep, with the epithet "rich in flocks"

---

[80] R. Osborne, *Classical Landscape with Figures*, 50-51.

[81] *BCH* 89, 1965, 665f.

[82] M. Jost, "The Distribution of Sanctuaries in Civic Space in Arkadia," in S. Alcock and R. Osborne, edd., *Placing the Gods. Sanctuaries and Sacred Space in Ancient Greece* (Oxford: Oxford University Press, 1994), 220-23. A. Chaniotis, "Habgierige Götter, habgierige Städte. Heiligtumsbesitz und Gebietsanspruch in den kretischen Staatsverträgen," *Ktema* 13 (1988): 21-39, postulates a similar role for mountainous shrines on Crete.

used from Homeric times forward.[83] And from the early eighth century BCE through the Classical period, Arkadian craftsmen celebrated the region's sheep production by creating small bronze sheep figurines of a quality and quantity not seen in other regions of Greece.[84] Indeed, sheep-production was so important and honorable among the Arkadians that wealthy men such as Praxiteles of Mantineia described themselves and their fortunes in terms of sheep.[85]

* * *   * * *   * * *

By highlighting some of the social, environmental, and economic variables that can shape the responses to animal management, this discussion has emphasized that each Greek community (indeed, each Greek) devised its own unique ways, methods, and goals for keeping animals to meet unique, social agendas. Since the rancher of Athens, absentee-stockman of Sparta, horse and cattle baron of Thessaly, and shepherding clan of Arkadia did not share similar goals, or similar physical environments, they did not share similar production methods. At Athens, the dry nature of the Attic countryside and the lack of year-round pasture prohibited the Athenians from raising horses and cattle in large numbers and ensured that sheep and goats, which could thrive on the scrub-covered hills, would predominate. But without the necessary socio-economic inducements, such as the export wool market and the unusually large state-sponsored demand for meat, animal husbandry in Athens would have remained a small-scale affair. Moreover, without the need for capital with which to perform socially necessary liturgies, the elite might not have pushed the limits of their natural resources. At Sparta helotage and sufficient natural grazing allowed the

---

[83] Pind. *Ol.* 6.100, 169; *Hymn. Hom.* 4.2; 18.2; *Il.* 2.605; Strabo 8.3.6. Simon. fr 104; Theoc. 22.157; Philostr. *VA* 8.7. Bacchyl. 11.95 calls it sheep-feeding Arkadia.

[84] See M. Voyatzis, *The Early Sanctuary of Athena Alea at Tegea. And Other Archaic Sanctuaries in Arcadia* (Göteborg: Paul Astrom, 1990).

[85] *IG* 5.2.47(1). See S. Hodkinson and H. Hodkinson, "Mantineia and the Mantinike: settlement and society in a Greek polis," *ABSA* 76 (1981): 271, 280.

Spartans to produce horses for cavalry or chariot-racing, as well as cattle, sheep, and goats for meat, cheese, hides, and wool. But without the social need to compete through gifts of cheese and meat or horse-racing and cavalry production, large herds would not have been necessary. In Thessaly, vast, well-watered pastures provided an unparalleled pastoral resource, but it was a social system, which encouraged elites to compete in producing huge herds of cattle, sheep, and goats, as well as the best racing stud, that allowed available pastures to be exploited. For the uplands, where many arable crops were difficult or impossible to grow, animals became a hedge resource and medium of exchange for the products of arable agriculture. In the end, no two communities, and no two individuals, raised animals in the same way or for the same reasons.

IV

THE PRESSURE FOR PASTURE
ANIMAL HUSBANDRY AND WAR

Sokrates: We shall have to cut out a piece of our neighbors' land, if we are to have enough to graze and to plant, and they in turn of ours if they too abandon themselves to the acquisition of wealth, disregarding the limit of what is necessary.

Glaukon: Inevitably, Sokrates.

Sokrates: We shall then go to war, Glaukon, shall we not?

Glaukon: Most certainly.

Plato, *Republic*, 373d-e

As we have seen, the number and even the breed of animals a Greek stockman could raise depended completely upon the quality and quantity of the grazing he could control. As a result, access to pasture weighed heavily on the minds of both large-scale and large-animal, producers, especially those who lived in regions deficient in permanent, unfarmable wetland resources such as Attika, southern Boiotia, the Megarid, and the northeastern Peloponnese. And even though the wealthy gentlemen of each community found their own local solutions to the problems of pasture, conflict over grazing rights, both between neighboring states and between neighboring citizens, was common to all. In the preface to his analysis of the great conflict between Athens and Sparta, Thucydides comments on the ubiquity and pettiness of border disputes during the Archaic and early Classical Periods.[1] It seems Sokrates was right: eventually everyone will need to "cut out a piece" of the neighbors' land in order to have enough to graze and plant. Because of this, elite desire for more pasture could affect even the poorest members of a community in a most life-altering way.

And yet, despite such conclusions by ancient authorities like Plato, scholars of ancient Greece do not discuss the impact of forcible pasture-acquisition, or armed pasture-protection on the social, political, and economic life of the ancient Greek community. For their part, the proponents of the agro-pastoralist/transhumance debate ignore questions of controlling pasture in their search for general animal management strategies.[2] Even broader studies of the ancient Greek countryside downplay the implications of pasture acquisition on intra- and inter-community life. Robin Osborne's *Classical Landscape with Figures*, for example, follows the functionalist arguments of P. Vidal-Naquet and M. Sartre and concludes that pastoral issues were insignificant to the ancients because the majority of the population was engaged in more settled, agricultural production.[3] Osborne argues that Greeks

---

[1] "By land no wars arose from which any considerable accession of power resulted; on the contrary, all those who had the same borders were against each other" (Thuc 1.15.2). Later in the work, when he discusses several of these disputes over borders in greater detail, Thucydides identifies the economic, specifically pastoral, interests that lay behind many of these conflicts. Thuc. 2.27 and 5.42. See discussion, below.

[2] S. Hodkinson, "Animal Husbandry in the Greek Polis," in C. R. Whittaker, ed., *Pastoral Economies of Ancient Greece and Rome* (Cambridge 1988) 35-74; and J. Skydsgaard, "Transhumance in Ancient Greece," in C. R. Whittaker, ed., *Pastoral Economies of Ancient Greece and Rome* (Cambridge: Cambridge University Press, 1988), 75-86. Though this is changing. C. Chandezon, *L'élevage en Grèce (fin V^e-fin Ier s. a. C.)* (Bordeaux: Ausonius, 2003), for example, is sensitive to temporal and regional variation and gives some consideration to war as a means to protect and acquire pasture, though he is mostly concerned with the epigraphic evidence from the Hellenistic period.

[3] Although Osborne is aware of problems involved with grazing and animal production, he generally agrees with the prevailing view that issues of pasture played a minor, mostly symbolic, role in the overall concerns of the

viewed their territory in terms of the world of the ephebe—the juvenile, uncivilized, largely pastoral, world of the frontier—and the world of the hoplite—the adult, inhabited, political, agricultural world of the settled community. As a result, pasture and pastoral issues were not an important part of the "adult" world of community decisions. Even Giovanna Daverio Rocchi, in her thorough study of the religious, political, and economic factors that shaped the ancient perception (and use) of the frontier, gives pasture-acquisition little detailed attention.[4] It is the same for studies of Archaic and Classical warfare.[5] W. R. Connor, for example, has suggested that Archaic Greek battles, though justified in the ancient sources as pragmatic grievances over border encroachments, were nothing more than symbolic articulations of sovereignty and identity.[6] Connor sees pre-Marathon hoplite battle as a social construct: Greek warfare was a ritualized display of honor, an expression of self and civic consciousness, a validation of social relationships both within and between *poleis*, advanced and negotiated under the guise of pragmatic grievances over border encroachments. Victor Davis Hanson, in his controversial study of the rise of the polis

---

Archaic and Classical polis. Robin Osborne, *Classical Landscape with Figures* (London: George Philip, 1987). P. Vidal-Naquet, "La tradition de l'hoplite athénien," in J.-P. Vernant, ed., *Problèmes de la guerre en Grèce ancienne* (Paris: La Haye, 1968), 161-181. M. Sartre, "Aspects économiques et aspects religieux de la frontière dans les cités grecques," *Ktema* 4 (1979): 213-24.

[4] G. Daverio Rocchi, *Frontiera e Confini nella Grecia Antica* (Roma: <<L'Erma>> di Bretscheider, 1988).

[5] E.g. V. D. Hanson, *The Western Way of War* (Berkeley and Los Angeles: University of California Press, 1989), and H. van Wees, *Greek Warfare. Myths and Realities* (London: Duckworth, 2004). Most studies on Archaic and Classical warfare tend to focus on the nature and origins of hoplite warfare and its relationship to arable agriculture. This is less true for studies of Hellenistic warfare. A. Chaniotis, *War in the Hellenistic World* (Malden, MA: Blackwell, 2005), for example, offers a sensitive treatment of the role agricultural and pastoral needs can play in war.

[6] W. R. Connor, "Early Greek Land Warfare as Symbolic Expression," *Past and Present* 119 (1988): 3-29.

and the hoplite agrarian revolution, agrees.[7] Hanson argues that even though Archaic and Classical hoplite-agriculturalists knew border lands were not significant to them in a purely agricultural respect, they valued these isolated tracts as symbols of territorial integrity and of group identity, reflective of the individual farmer's own endless haggling with his neighbors over property boundaries. In effect, the hoplites of the Archaic and Classical periods conceptualized the entire polis as their own farms writ large and would allow no outsider to trespass. Yet it seems out of character for the doughty hoplite-agriculturalists, whose pragmatic way of life Hanson takes such pains to describe, to leave their beloved farms and fight for land that was of little or no use to them. Even Hanson is puzzled by this apparent contradiction: "it is hard to see much real profit in a hoplite army's acquisition of a neighbor's scrub land on the borders."[8] There must be more to the picture.

While it is clear that early hoplite warfare did bring about the sense of community identity and regional sovereignty Connor postulates, it is dangerous to dismiss or ignore the pragmatic *casus belli* of early battles that the ancients themselves asserted so blithely. In order to work as *casus belli*, to function as justifications for killing the neighbors and taking their land, borders had to represent something tangible and potentially useful to those fighting. Models such as Connor's analyze the psychological results of hoplite battle, the socio-political dynamic in play, but they sidestep the real-life connections these bloody conflicts had to the men who fought them, the "hows and whys" pragmatic border grievances could serve so effectively, and for so long, as symbols for sovereignty and individual identity. Even Hanson's agricultural symbolism

---

[7] V. D. Hanson, *The Other Greeks. The Family Farm and the Roots of Western Civilization* (Berkeley and Los Angeles: University of California Press, 1999).

[8] Hanson, *The Other Greeks*, 251. See V. D. Hanson, "Hoplite battle as ancient Greek warfare: when, where, and why?," in H. van Wees, ed., *War and Violence in Ancient Greece* (London: Duckworth, 2000), 201-232, for a reiteration of this sentiment.

does not fully explain why middle-class, arable-farming hoplites went to war to protect distant and useless tracts of land, far from their own farms.

But perhaps the contradiction that has puzzled Hanson, i.e. single-minded agriculturalists fighting over agriculturally insignificant borderland, stems from a fundamental misinterpretation of the use of this borderland. Perhaps the identity expressed and affirmed by means of these border disputes was elite, not middle class— elite identity that was constructed on a foundation of animal wealth. The "scrub-lands" of the Greek countryside may have been insignificant to arable farmers but they were quite useful and even essential to large-scale sheep and goat producers. The rocky border regions between Attika and Boiotia, for example, supported an impressive array of plant-life that could feed numerous herds and flocks.[9] And naturally irrigated river valleys and marshy coastal plains, such as those between Chalkis and Eretria and Argos and Sparta, were excellent for keeping horses and cattle. By focusing attention on the pastoral resources of borderlands, it is possible to understand the motives and pretexts behind the border wars of the Archaic and Classical periods and consequently unlock the symbolism Connor and others identify.

We begin with the earliest and the least known Greek war, the First Messenian War, the war that pulled Sparta into lasting conflict with the inhabitants of Messenia.[10] Although there are no contemporary accounts, later sources all agree that hostilities came about because the Spartans had appropriated an upland river valley on the

---

[9] O. Rackham, "Observations on the Historical Ecology of Boeotia," *ABSA* 78 (1983): 291-351.

[10] The traditions underlying the First Messenian War have increasingly come into question, with some scholars dismissing the entire event as later invention. See N. Luraghi, "The Imaginary conquest of the Helots," in N. Luraghi and S. E. Alcock, edd., *Helots and their masters in Laconia and Messenia: histories, ideologies, structures* (Cambridge, MA: Harvard University Press, 2003), 109-141, and P. Cartledge, "Raising hell? The Helot Mirage—a personal re-view," in N. Luraghi and S. E. Alcock, edd., *Helots and their masters in Laconia and Messenia: histories, ideologies, structures* (Cambridge, MA: Harvard University Press, 2003), 12-30.

western slope of the Taygetos range. Why? What were the
Spartans doing on the Messenian side of Taygetos and why
did they want a well-watered upland valley? In his
commentary on Pausanias, J. G. Frazer, located the
disputed land near modern Volimnos, in a naturally
irrigated region along a tributary of the Nedon River
known for its transhumant herds of sheep and goats.[11]
According to tradition, once the Spartans were in
possession of the land, they set up a shrine to Artemis
Limnatis, "Artemis of the Wetlands," a poorly understood
goddess seemingly connected with shepherds and
herding.[12] Now this particular Artemis was fairly unique to
the Spartans, worshipped primarily in marshy regions of
Lakonian territory, or marshy places influenced by Spartan
culture. Her chief shrine was located in the center of Sparta
itself, near the Eurotas River, probably in or nearby the
district of Limnai; others were found in Spartan-controlled
Epidauros Limera, in a wetlands area along the sea, and
near Arkadian Tegea, situated along the main road between
Sparta and Tegea, next to a river flowing into the Tegean
marshlands, which is still used today by sheep and cattle
graziers.[13] The only attested shrines outside of Spartan
control, at Patrai and near Sikyon, also seem to have had
some connection with springs, water, and pasture;
Pausanias, for example, locates the Sikyonian temple near a
spring and shepherd camp called the "Dripping Cave."[14]

A closer link to grazing can be seen in the most
dramatic and extensive of all archaic Greek border wars, the

[11] J. G. Frazer, *Pausanias's Description of Greece*, (London: Biblo and
Tannen, 1898), 3.427, 6. E. C. Semple, "The Influence of Geographic
Conditions upon Ancient Mediterranean Stock-Raising," *Annals of the
Association of American Geographers* 12 (1922): 3-38.

[12] Tacitus *Annales* 4.43; Pausanias 3.7.4; 4.4.2; 4.31.3; Diodoros 8.7. Frazer
locates the place of the dispute by means of an inscription from the shrine of
Artemis of the Wetlands. See P. Cartledge, *Sparta and Lakonia. A regional
history 1300-362 BC* (London: Routledge, 1979), 112-13, for discussion of the
shrine and its historical context.

[13] Limnai (Paus. 3.14.2); Epidauros Limera (Paus. 3.23.10); Tegea (Paus.
8.53.11). S. Hodkinson and H. Hodkinson, "Mantineia and the Mantinike:
settlement and society in a Greek polis," *ABSA* 76 (1981): 239-96.

[14] Patrai (Paus. 1.20.7); Sikyon (Paus. 2.7.6).

seventh century dispute between the Chalkidians and Eretrians over the coastal plain of the Lelas River, the so-called Lelantine War.[15] In time, this local border dispute swelled to become a pan-Hellenic war as each side called upon friends and allies, some from as far afield as Samos and Miletos.[16] The extraordinarily wide connections of the original combatants and the ultimate scale of the war (rather than the reasons behind the original dispute) attracted the attention of Thucydides and have continued to hold modern scholars.[17] As a result, the origins of the war have received less attention than the complex relationships between the various participants. And yet the motive to fight is fairly clear: the Chalkidians and Eretrians went to war for control of the Lelantine plain.

Why? What made this land so valuable? The Lelantine plain was a treeless, well-watered coastal region, and as such it offered the one area of superb horse pasture on an otherwise precipitous and rocky island. Both the Chalkidians and Eretrians were renown horse-breeders, and horses seem to have played an important role in constructing Euboian elite identity—the ruling class of Chalkis called themselves the *Hippobotai*, and those of Eretria referred to themselves as the *Hippeis*. Moreover, the sources tell us that the Lelantine war was fought primarily with cavalry.[18] Both sides wanted to control the grazing resources of the plain in order to increase their

---

[15] Thuc. 1.15; cf. Strabo 10.1.12; Plut. *Moralia* 153 F.

[16] Even though the scale of participation during the Lelantine War was unprecedented, this conflict between the Chalkidians and Etretrians over the Lelantine Plain was neither new nor decisive. A century or more later the Megarian poet Theognis observes that the Lelantine Plain has once again been laid waste by war (Theog. 892). See Boardman, "Early Euboean Pottery and History," *ABSA* 52 (1957): 5ff., argues that the Lelantine War dragged out for at least a century.

[17] E.g. V. Parker, *Untersuchen zum Lalantischen Krieg und verwandten Problemen der frühgriechischen Geschichte* (Stuttgart: Franz Steiner Verlag, 1997).

[18] Arist. *Pol.* 1289b.38. This appears to have been primarily an aristocratic conflict, for both sides banned projectile weapons such as spear, slings or arrows—the sorts of weapons common to less well-to-do soldiers. For the ban on weapons see Strabo 10.1.12 and Archilochos Fragment 3 (West).

herds and thus their cavalry strength and elite status at the expense of their neighbor. Lasting, pan-Hellenic war was the result.

While the Lelantine War was fought over a fertile, relatively flat and well-watered region, desirable for both grazing and farming, other disputes centered around less fertile borders that could be useful only for pasture, such as Panakton, between Athens and Thebes. According to Thucydides, the Boiotians destroyed the Athenian frontier community of Panakton in 420 BCE because the Athenians had defied an ancient pact. The Boiotians claimed that Athens and Thebes had fought over the border for years, but since battle could not affect a lasting solution, both exchanged oaths not to inhabit the district and to graze it in common.[19] By fortifying and inhabiting Panakton, the Athenians had violated the oath. Though Thucydides does not comment on the legitimacy of the Boiotian claim (he merely reports what the Boiotians assert), parallel border agreements from other regions of Greece suggest that such arrangements were not uncommon.[20] Around 369 BCE, for example, Arkadian Orchomenos and its neighbor Methydrion agreed to set up a communal grazing ground along their shared, and much disputed, border, swearing oaths allowing the citizens from both communities to use the frontier pastures.[21]

Unfortunately, the ancient narratives dealing with the disputes over Panakton and the Lelantine Plain are compressed and consequently obscure much of the complexity behind the claims and uses of these lands. The

[19] Since both the Athenians and Boiotians originally agreed to graze the land in common, the original motive behind the ancient war must have been control of the region's pasture resources. For a review of the sources on this ancient dispute see M. Munn, "New Light on Panakton and the Attic-Boiotian Frontier," in H. Beister and J. Buckler, edd., Boiotika. Vortäge vom 5. Internationalen Böotien-Kolloquium zu Ehren von Professor Dr. Siegfried Lauffer (München: Editio Maris, 1989), 231-244. On the archaeological evidence for the history of Panakton see M. Munn, "The First Excavations at Panakton on the Attic-Boiotian Frontier," Boeotia Antiqua 6 (1996): 47-58.

[20] See Chandezon, L'élevage en Grèce, 331-390, for discussion.

[21] BCH 39 (1915): 55. See Daverio-Rocchi, Frontiera e Confini nella Grecia Antica, 96-99.

fourth-century BCE Oxyrhynchos Historian, however, is much more explicit, and his description of a border war between the Western Lokrians and Phokians offers some essential context for the issues discussed so far:

> There is a piece of land near Mt. Parnassos, which is disputed by the Phokians and the Ozolian Lokrians. They had fought about it even before this. Each side uses this land as pasture, and whatever side happens to find the other doing this collects a large force and carries off the herds.[22] Many such disputes had occurred before, and they always resolved them for the most part by discussion and judicial proceedings, one with another, but on this occasion the Lokrians made a counter-raid to snatch back herds in return for those the Phokians had run off with, and the Phokians immediately invaded Lokris. As their land was being ravaged, the Lokrians sent ambassadors to the Boiotians, accusing the Phokians and saying that the Boiotians ought to help them.... When the Phokians heard the news from Boiotia they retreated from Lokris again and sent ambassadors to Sparta immediately, saying that the Spartans should warn the Boiotians not to enter Phokis.[23]

Here, there can be no mistake about motive: the Phokians and Lokrians came to blows over a border pasture both claimed and both grazed. As with the Lelantine Plain and Panakton, this dispute was long-standing; the author states that these communities had fought over the pasture for some time, although formal battle was rare and raiding and arbitration characterized normal relations.[24] This type of compromise is probably behind the oath sworn between the Athenians and Thebans at Panakton; and the constant raiding and skirmishing echoes the exchanges between Chalkis and Etretria before conflict escalated.[25] What is

---

[22] This is the sort of incessant raiding Thucydides means in 1.5-6.

[23] *Hell. Oxy.* 18.3-4 (Bartoletti) = Chambers 21.3-4.

[24] Cf. *Il* 9.404-5: "In Phoibos Apollo's house on Pytho's sheer cliffs, cattle and fat sheep can all be had for the raiding."

[25] A similar situation seems to have occurred in the Thyreatis, a border territory between Argos and Sparta in the northeastern Peloponnese, that contained both marshy coastal pasture for horses and cattle and the high

unusual, however, about the Phokian-Lokrian dispute, and perhaps adds further context to the build-up to the Lelantine War, is that on this occasion each side did not end the dispute in the traditional way, through battle or arbitration. Instead, the Lokrians called in their allies the Boiotians, and the Phokians called in their allies the Spartans. And in time this inconsequential border conflict over grazing and sheep escalated into a pan-Hellenic war (the Korinthian War of 395-387 BCE), involving all the major Greek cities of the mainland. While the tensions in Greece between the "Great Powers" happened to be heightened enough to spin the Lokrian-Phokian dispute into a pan-Hellenic event, this local dispute gone wrong emphatically illustrates how local pressure for pasture can affect the life and health of communities, at all levels, even for those not connected with animal husbandry or the border pasture.

---

meadows in the Parnon range for sheep and goats. According to Herodotos, around 550 BCE the Spartans cut out the Thyreatis from the rest of Argive territory and occupied it (Hdt. 1.82; cf. *Anth. Pal.* 7.244; 7.431-2). Upon learning of this outrage, the Argives mobilized their army and attempted to oust the Spartans. They failed: even after a series of battles, including one in which only 300 select men from each side fought, the Argives were unable to evict the Spartan invaders. From this time forward, the Argives vowed to avenge their dead and regain their territory. Unfortunately, they were more persistent than successful, skirmishing along the border as well as pursuing the matter through diplomacy whenever they felt Sparta's control weaken. In 420 BCE, for example, the Argives raised the issue of Thyrea when the Spartans asked them for a 50-year truce. According to Thucydides, the Argives spoke for some time about the circumstances surrounding the original theft and their inability to regain their territory through force of arms. In the end, although the Argives considered submitting the matter to outside arbitration, they decided to settle the conflict as they had always done in the past, by force of arms. And just as in 550, the Spartans were victorious, retaining control of the coveted region throughout the fifth and fourth centuries. In fact, the Argives did not regain Thyrea until 338 BCE, when Philip of Macedon returned it to them as part of his general settlement for Greece (Paus. 2.38.5; Polyb. 9.28.7; 18.14.7). Euripides, *Electra* 413, observed that shepherds commonly tended flocks of sheep and goats in the Parnon region. Thucydides, 2.27.2, reports that the Spartans especially valued the region for grazing and for this reason they settled Aiginetan refugees in the district of Thyrea (431 BCE) to inhabit and to graze it. H. A. Koster, *The Ecology of Pastoralism in relation to changing patterns of Land Use in the Northwestern Peloponnese* (Diss., University of Pennsylvania,1976), has documented the value of the Thyreatis region among modern pastoralists.

The peculiar rhythms of Greek pasture war are best seen in a decree from the Hellenistic period concerning a goat pasture in the northeastern Peloponnese, claimed by both Epidauros and Hermione.[26] Since the early Archaic period, this rocky, brush covered border had been carefully marked with inscribed boundary stones.[27] Early in their history both communities had invested time and resources in defining the ownership of these agriculturally insignificant hills and valleys, for on such productive grazing fortunes could be made—one mid-fourth century BCE stockman named Timainetos was so successful at goat production in this region that he was able to dedicate a small altar at Epidauros from "the tithe of the goats."[28] As their herds and prosperity grew conflict erupted, and each side sought sole ownership of the pasture. In the end, though, neither side could fully dominate the other and, as we have seen elsewhere, compromise was reached: both communities agreed to share the pastures equally. But as at Panakton, the sharing did not last; it seems that substantial citizens like Timainetos wanted to have sole control over this productive sheep and goat grazing. And even though war again did not achieve a lasting solution, and seems never to have been full-scale, both sides refused to end hostilities. In fact, so indecisive and regionally disruptive did the on-and-off pasture war become that by the second century BCE both Epidauros and Hermione were forced into arbitration by a higher authority, the Achaian League.[29] After reviewing the case, however, the commission from

[26] IG 4^2.1.75.

[27] See M. H. Jameson, C. N. Runnels, T H. van Andel, edd., *A Greek Countryside. The Southern Argolid from Prehistory to the Present Day* (Stanford: Stanford University Press, 1994), 603-605. The Delphians erected a similar border demarcation between their own highland pasture and that of the Ambryssians and Phlygonians (FD 3.2.140-7). See Osborne, *Classical Landscape with Figures*, 50, for further discussion.

[28] SEG 26.451. See Koster, *The Ecology of Pastoralism*, 186, for a discussion of the pastoral resources of the region.

[29] M. Dixon, "IG IV2.1.75 and the Date of the Arbitration between Epidauros and Hermion," ZPE 137 (2001): 169-173. See IG 4^2.1.70, 71=Syll^3 471, for a similar border region disputed between Epidauros and Korinth, and also IG 4^2.1.72, between Arsinoë (Methana) and Epidauros. See Osborne,

the League decided that the earlier compromise should stand: the land would remain in common use, with each side now fined for misconduct.

The decision of the League officials—their recognition of the *status quo* that the goat pasture remain in common use—seems to have been the end result of most border conflicts by the late Classical and early Hellenistic periods. But as the Epidauros-Hermione dispute shows this common use was secure *only* if a higher power guaranteed or advocated it—in this case the Achaian League. The threat of appropriation always lurked in the background, as fortunes and ambitions of the communities and individuals involved waxed and waned. As Sokrates warned Glaukon, the pressure for pasture would break out in war, war that could disrupt the life and health of entire regions. For this reason Philip of Macedon wished to settle all border disputes between the Greeks, and even forbade communities from fighting each other.[30] This ultimately proved unsuccessful, but it is significant that the federated regional governments of the Hellenistic Age tried to limit in whatever way they could the collateral damaged caused by communities and individuals seeking more grazing.[31]

But limiting collateral damage and ensuring access to scarce regional pasture resources was not just an invention of Hellenistic federalism or Hellenistic monarchs. Beginning in the early Archaic, dedication to a god offered a safe way to protect pasture and even *create* grazing in fertile, naturally irrigated areas that would otherwise have been taken for farming. In many respects, creating "sacred" lands offered a series of unique solutions to the problems

---

Classical Landscape with Figures, 162-64, for comments on the former, C. Mee and H. A. Forbes, *A Rough and Rocky Place*, 74f., for the latter.

[30] At this time Philip settled many border disputes by returning lands to their "original" owners. For example, Tegea regained Siritis (Theopompos F 238); Megalopolis Belemina (Livy 38.44; Polyb. 9.28; Paus. 8.35.4); Messenia Denthalia (Tac.*An.* 4.43.3) and Argos Thyrea (Paus. 2.38.5; Polyb. 9.28.7; 18.14.7). For a discussion of Thyrea's role in Philip's decisions see R. A. Tomlinson, *Argos and the Argolid* (Ithaca, NY: Cornell University Press, 1972), 145-146.

[31] Chandezon, *L'élevage en Grèce (fin V^e-fin Ier s. a. C)*, 309-350.

caused by communal grazing, border territory, and interstate war. And there were benefits to sacred lands that went beyond simply protecting grazing and dampening conflict. By placing disputed pastures under the protection of a god, animal producers could restrict use through a system of fines and penalties, guard against over-grazing, ensure a local and lucrative supply of sacrificial victims, and even guarantee camping and grazing facilities for religious visitors to use while attending local festivals and ceremonies. But with unique solutions come unique problems: as we saw in the case of Epidauros and Hermione, not all neighbors want to share land, and in the case of sacred lands, not all want good, fertile land to be unfarmed and under the control of a religious institution. Sacred lands tended to inspire sacred wars—if individuals wanted land badly enough, they would "cut it out" from their neighbor, even if he was a god.

The famous Sacred Land of Apollo near Delphi best illustrates the strategies behind creating sacred pastures and underscores the point that even divine protection did not guarantee safety from encroachment or seizure. From its creation in 590 BCE (as a result of the First Sacred War) the Sacred Land of Apollo saw at least three major pan-Hellenic sacred wars fought in order to preserve its communal and unfarmed pastures. I have suggested elsewhere that regional war played a pivotal role in creating and maintaining a communal pasture to serve the needs of the oracular sanctuary of Apollo at Delphi, and we do not need to revisit all of those arguments here.[32] A short overview of the characteristics of the Sacred Land will suffice.

The Sacred Land of Apollo lies below the sanctuary of Delphi, in the fertile, naturally irrigated Pleistos River valley. The land had been the sovereign territory of the community of Krisa until around 590 BCE, when the Krisaians, who supplied the sacrificial animals to Delphi, began charging exorbitant prices for animals and otherwise

[32] T. Howe, "Pastoralism, the Delphic Amphiktyony and the First Sacred War: the creation of Apollo's sacred pastures," *Historia* 52 (2003) 129-46.

abusing visitors. A coalition of Greeks, mostly from central
Greece, banded together and agreed to end Krisa's reign of
terror by means of a holy war. Once victorious, the
coalition sanctified the Krisaisans' land to Apollo and
celebrated a new and improved Pythian festival in his
honor, complete with expanded chariot and horse races,
now possible because of the newly confiscated Krisaian
Plain.[33] In order to prohibit the future abuses such as those
perpetrated by Krisa, the coalition agreed that an
international council, the Amphictyony of Anthela and
Delphi, composed of representatives from all of the major
communities in Greece, would from this time forward
administer the oracular shrine and adjoining Sacred Land.[34]
All members of the Amphiktyony had to swear an oath
that they would protect the land from farming, habitation,
and overgrazing. The Krisaian Plain would supply grazing
only for the god's own animals and those belonging to
short-term visitors. The council imposed a 30-day limit on
guest camping and pasturing. And in order to ensure that
people did not overstay their welcome, the officials of the
Amphiktyony, the *hieromnemones*, patrolled the land and
identified any abusers, who were then fined by the council.
If the *hieromnemones* did not patrol the land regularly,
they were themselves fined. If the abusers failed to pay the
fine, or began farming or building on the god's land, they
would be removed by force.[35]

This practice of allowing festal participants to graze
their animals on sacred land seems to have been fairly
widespread in Greece. Xenophon, for example, ensured that
the territory he dedicated to Artemis at Skillous contained
"a meadow and hills covered with trees—suitable for

---

[33] See U. Kahrstedt, "Delphi und das heilige Land des Apollon," in G. E.
Mylonas and D. Raymond, edd., *Studies presented to David Moore Robinson II*
(St. Louis: Washington University Press,1953), 749-757, for a discussion of the
extent of the sacred land and its independent status with respect to the *polis* of
Delphi.

[34] See G. Roux, *L'Amphictionie, Delphes et le temple d'Apollo au VIe
siècle* (Lyons: Maison de l'Orient, 1979), 1-59, for a discussion of the
Amphiktyony and its responsibilities.

[35] *Syll*^3 145.15-25 = *CID* i.10.15-26.

raising pigs, goats and cattle and horses, so that even the beasts of burden belonging to those who attend the festival might be well fed."[36] Likewise, in fourth-century Arkadian Tegea, at the sanctuary of Athena Alea, foreigners and citizens were permitted to pasture both dedicated and personal animals on sacred land for the duration of the celebrations, so long as they had, in fact, come for the festival. In addition, the officials of the sanctuary were admonished to patrol the sacred territory and fine those who had outstayed their welcome. If the officials were remiss, they were themselves fined, just like the Delphic *hieromnemones*.[37] The purpose for all of this was more than just camping and grazing facilities for visitors. The Sacred Land was a profit-oriented business. The *hieromnemones* protected the Sacred Land so that the god's own sacred animals might be safe from theft and lack of food through overgrazing. The Sacred Land allowed Apollo to breed his own sacrificial animal to sell to pilgrims.[38] This was the primary motive for sanctifying the Krisaian plain in the first place.[39] And Delphian Apollo was not unusual in supplying his own vicitims. Many sanctuaries kept temple herds on temple lands, which the officials would regularly sell to suppliants.[40] In the end, sacred lands were not only a public grazing area on which elite visitors displayed their dedications and pastured their beasts of

---

[36] *Anabasis* 5.3.11-12

[37] *IG* V. 2, 3.1-21.

[38] By the second century BCE, Apollo owned huge sacred herds, which were pastured on the Sacred Land and sold on behalf of the sanctuary. It is even possible that, in addition to sacrificial animals, the guardians of Delphi were producing racers for the chariot-racing trade. Syll^3 636; Syll^3 826 G.

[39] The archaic *Homeric Hymn to Apollo* clearly implies (533f.) that the god would no longer lack sacrificial animals.

[40] The herds of Hera Lacinia in southern Italy were famed for their numbers and the wealth they brought the temple (Livy 24.3; Strabo 6.1.11; Cic. *de invent.* 2.1). Athena Alea in Tegea also had temple herds, which grazed on sacred land near the temple. *IG* v. 2.3.1-21. See also *ID* 503.23-4 for the sacred herds of Apollo on Delos, and especially *IG* II/III 1639.15-16, 17; 1638.66; 1640.28, for the sale of wool from sacred sheep.

burden, but also a money-making concern for a sanctuary.[41]

But what if fines, patrols, and even religious sanctions were not enough to protect the god's land and the sacred animals? The god would have to declare war. Ultimately, overseeing bodies such as the Amphiktyony of Anthela and Delphi were forced to declare sacred war against those who cultivated, inhabited, or overgrazed the god's pastures. The Third Sacred War (356-346 BCE) illustrates the process well.[42] In 357 the Delphic *hieromnemones* discovered neighboring Phokians cultivating the Sacred Land where it bordered the sovereign territory of Phokis and imposed a fine of "a large number of talents."[43] In response, the Phokians claimed the fine was unjust and refused to abandon their newly won land. The *hieromnemones* ordered the Phokians to pay the fine or have their land cursed.[44] This threat pushed Philomelos, leader of the Phokians, to seize all of the Sacred Land and the sanctuary at Delphi. And it was these actions, coupled with the already committed sacrilege against the Sacred Land, that compelled the Amphiktyony to declare a sacred war against the Phokis. The Amphiktyony searched central Greece for a leader and soon settled on Philip of Macedon, who was ready for any excuse to move south and meddle in the affairs of the southern Greeks. As *hegemon* of the sacred army, Philip defeated the Phokians (after a few setbacks,

---

[41] I am grateful to Jeremy McInerney for the suggestion that the distinction between public and private herds on sacred land may be blurred. He suggested to me that a close reading of the land leases from the sanctuary of Artemis at nearby Hyampolis (*IG* 9.1.87) and Pausanias (10.35) may suggest that tenants of the sacred land of Artemis offered animals to the goddess in return for the use of the land. In this way the sanctuary received a supply of ready victims for sale to suppliants.

[42] J. Buckler, "Thebes, Delphoi, and the Outbreak of the Third Sacred War," in P. Roesch and G. Argoud, edd., *La Béotie antique* (Paris: Éditions du Centre national de la recherché scientifique, 1985), 237-246, and *Philip II and the sacred War* (Leiden: E. J. Brill, 1989), 17ff.

[43] Diod. 16.23.3.

[44] Diod 16.23.5. This account of the events follows the chronology of N.G.L. Hammond, "Diodorus' narrative of the sacred war," *JHS* 57 (1937): 44-77.

such as the humiliating defeat at Crocus Field), stripped them of their seat on the Amphiktyony, and enforced payment of the war indemnity. The Third Sacred War, a war begun over pasture, allowed Philip the entry he needed into southern Greek politics.

The events of the Fourth Sacred War show in an even more dramatic fashion how pasture abuse on the sacred land could spin out of control into regional war. Unlike the Third Sacred War, in which hostilities were initiated because the Phokians had cultivated the land *and* seized the oracular shrine, the Fourth Sacred War was declared because another neighbor of the Sacred Land, the Amphissans, refused to cease their unlawful grazing and cultivation of the sacred land.[45] As protector of Delphi, Philip was again called in. But not all southern Greeks welcomed this. Demosthenes in particular suspected Philip's motives and the need for another sacred war. After much debate with his rival Aischines, he convinced the Athenians, and ultimately the Thebans, that they should oppose Philip. The Battle of Chaironeia in 338 BCE was the result. Here, Philip and his son Alexander defeated a coalition of Greeks and in the process, ended the age of the independent polis. In the next year, Philip marched to Corinth, imposed the so-called League of Corinth on the Greeks and forbade them to fight among themselves over borders. It seems that Philip's experience with two sacred wars over pasture had convinced him that Greeks could not remain in their borders unless a more powerful entity forced them to do so.

Sacred land, sacred war, and pressure for pasture were serious business and more common than is usually supposed. In fact, sacred war over pasture seems to lay behind the declaration of the so-called Great Peloponnesian war between Athens and Sparta. In his background to the

---

[45] Dem. 18.154-155. Aischines 3.129. For a discussion of the Fourth Sacred War see R. Sealey, *A History of the Greek City States 700-338 B.C.* (Berkeley and Los Angeles: University of California Press, 1976), 484-491, and J. R. Ellis, *Philip II and Macedonian Imperialism* (London: Thames and Hudson, 1976), 186-198.

war, Thucydides lists the Corcyrean affair, Potidaia, and
the Megarian Decree as the decisive events that brought
Sparta and Athens to blows. Of these, the final straw was
the Megarian Decree. The reasons for issuing the Decree
are complex and revolve around Megarian appropriation of
land. Tensions heightened in 432 BCE, when Perikles
formally accused the Megarians of cultivating the *Hiera
Orgas* and threatened to close all Athenian ports to Megara
(the Megarian Decree) if the Orgas was not returned to its
rightful use.[46] This Hiera Orgas, or "Holy meadow," was
located in a border region, next to the regionally important
sanctuary of Demeter and Kore at Eleusis, and set aside for
grazing—cultivation was expressly prohibited.[47] In 432, the
Megarians refused to remove their settlements on the
Orgas and in response Perikles proposed to the Athenian
Ekklesia that a herald be sent to both the Megarians and
the Spartans to denounce Megara. This was done, though it
seems to have had little effect, since the continued violation
of the Orgas by Megara prompted the Athenians to close
their ports. At least that is what Thucydides 1.139.2 argues:

> But the Athenians would not heed their other demands
> and refused to rescind the decree prohibiting the
> Megarians from using the ports of the Athenian empire
> and the Attic market, saying in addition that *the
> Megarians were cultivating the sacred land and the
> borderland not marked by boundary stones* and also
> harboring runaway slaves.

Although the Athenians employed the Megarian
offense against the Orgas and other borderland as rhetoric
for closing their ports, this might not have been so much
hot air.[48] As we have seen, well-watered, open grazing

---

[46] Plutarch (*Perikles* 30.2), says that Perikles "publicly accused them of
parceling off the Hiera Orgas." The verb *apotemno* used here in middle voice
suggests subdivision for the purpose of profitable use.

[47] Perhaps just as both Thebans and Athenians used the Panakton region.
S. van de Maele, "L' Orgas Eleusinienne: Etude Topographique," *Melanges
Edouard Delabeque* (Marseilles,1983) 417-433, identifies the Botsika plain with
the *Hiera Orgas*.

[48] H. Bowden, *Classical Athens and the Delphic Oracle. Divination and
Democracy* (Cambridge: Cambridge University Press, 2005) is the most recent

areas such as the Orgas (which after all means "meadow," and implies moist, fertile land) and the Megarid-Attic border were much in demand among the Athenian animal producers, who were becoming rich because of the state-supported sacrificial market.[49] Moreover, for Perikles' pretext to make any sense as a pretext, it had to have some intrinsic value to the audience. Perikles had to pick a reason that resonated with the demos, or at least the leaders of the demos, if he were to convince them to support the Megarian Decree. He picks two, that appeal to rich and poor in their own ways: the seizure and cultivation of Hiera Orgas and the borderland, and the harboring of runaway slaves. The appropriation by the Megarians of these important grazing resource would certainly have stirred up wealthy pastoralists against the Megarians, and the symbolic attack on a sanctuary would have wide appeal in Athens, especially among the lower classes. Likewise, harboring fugitive slaves would give all Athenians pause and thus induce them to support the decree, and, consequently, the war. In the end, the inclusion of the Orgas among the Athenian reasons for continuing the Megarian Decree suggests that pastoral politics, such as those employed in the Delphic Sacred Wars were at work on some level.

Indeed, pasture continued to define the fate of the Hiera Orgas throughout the Classical Period. Almost fifty years after the conclusion of the Great Peloponnesian War, in 352/1 BCE, the Megarians again seized and parceled the Orgas out for farming. In response, as [Demosthenes] 13.32-3, tells us, the Athenian Ekklesia decreed that the army mobilize and prevent the Megarians from subdividing the Orgas. This time, however, the cultivation of the scared land did not result in war, or not a prolonged one, at any rate. A resolution of the Athenian Ekklesia

---

scholar to downplay the importance of Megarian encroachment on the Hiera Orgas in the Athenian decision to issue the Megarian Degree and consequently take steps towards war.

[49] See discussion in Chapter III, above.

gives the substance of the decree to which Demosthenes refers: The Athenians went out with force of arms, expelled the Megarians, and appointed one of the *strategoi* (the board of 10 generals) with the job of guarding the Orgas against further abuse.[50] The Athenians then sent to Delphi to ask what should be done with the Orgas, whether they should hire out for farming the areas now under cultivation, in order to build a new pavilion for Demeter and Kore, or return the land to the earlier, uncultivated status.[51] Didymus, in his commentary on [Demosthenes] 13.23, provides the oracle's decision: "the god had answered that it was most desirable and best not to cultivate the land." He also adds that the Megarians agreed to the decision of the oracle and to the boundaries set by the priests of Demeter and Kore, Lakrateides and his successor Hierokleides.

The similarities between the Hiera Orgas and the Sacred land of Apollo during the Classical period are striking: a high official is detached to guard the land from encroachment, one of the 10 *strategoi*, in the case of the Orgas, the entire body of *hieromnemones* of the Delphic Amphiktyony, in the case of the Sacred Land. These guards act as a deterrent, but when encroachment ultimately occurs, war is the result. It is in this context that we must view Perikles' justifications for issuing the Megarian Decree, and not be so quick to dismiss as mere pretext both the symbolic and practical role sacred lands such as the Hiera Orgas and the Sacred Land of Apollo played in Ancient Greece, especially in the charged inter-state rivalries of the Classical period. Both sanctuaries and animal producers needed pastures like the Orgas, and politicians were not above using this need to further political and territorial ambitions. In the end, we should be in no way surprised that the Delphic Oracle told the Athenians in 351 to return the Hiera Orgas to its unfarmed

---

[50] *IG* II² 204. Cf. *Insc.Eryth* 1.17, for guardians appointed over pasture.

[51] *IG* II² 204. ll. 23-30.

status. After all, as a proponent of pastoral politics, what else could the oracle do?

***   ***   ***

In ancient Greece pasture was an all-important commodity. Many of the most famous wars of the Archaic and Classical periods were fought for control of border pastures. While the majority of these pasture disputes often entailed nothing more serious than border skirmishing, they could, under the right conditions, devolve into full-scale regional war, such as the four Sacred Wars, the Lelantine War, the Korinthian War, and even the Peloponnesian War. Yet for all the energy expended, these border conflicts rarely resulted in any lasting solutions; the Lokrians and Phokians, for example, had been quarreling over the pastures of Parnassos for centuries and even the Korinthian War did not end the dispute, as the Third and Fourth Sacred Wars illustrate. At best, pasture wars ended in a draw, with the land given over, albeit grudgingly, to common or religious use, in some cases with grazing fees imposed on both sides. Such a solution is not all that surprising since pasturage issues, connected as they were to factors of both economics and prestige, could never be settled once and for all. Fortunes and ambitions would come and go as economic and social conditions shifted. As Sokrates wisely observed, pressure for pasture inevitably results in someone cutting out land from the neighbors.

V

THE POLITICS OF DISPLAY
ANIMALS, IDENTITY AND POWER

Athenians, I [Alkibiades] have a better right to command
than others...by reason of the magnificence with which I
represented it [Athens] at the Olympic Games when I
sent into the lists seven chariots, a number never before
entered by any private person, and won the first prize,
and was second and fourth, and took care to have
everything else in a style worthy of my victory. Custom
regards such displays as honorable and they cannot be
made without leaving behind them an impression of
power.

Thucydides, 6.16.1-4

Now when the Pythian festival was approaching, Jason
sent orders to his cities to make ready cattle, sheep, goats
and swine for sacrifice. And it was said that although he
laid upon each city a very moderate demand, there were
contributed no fewer than a thousand cattle and more
than ten thousand other animals...for he was intending,
it was said, to be himself the director both of the festal
assembly in honor of the gods and of the games.

Xenophon, *Hellenica*, 6.4.29

Among the ancient Greeks, elite social relations were
essentially evaluative and competitive, with the result that
public performances such as sport, drama, and ritual were
not mere entertainment but distinctive arenas in which
values, status relationships and other cultural signifiers

could be formulated and reformulated.[1] Animals played an essential role in these negotiations. For example, Alkibiades' horse and chariot victories at the Olympic games of 415 BCE elicited more than praise or congratulations from his fellow Athenians. By outstripping the competitors from other Greek cities, Alkibiades brought honor both to himself and to his community, since victory at such an important pan-Hellenic religious festival was seen as a manifestation of heroic ability as well as divine support. As Leslie Kurke has shown, victory at the crown games bestowed a certain talismanic power on both the victor and his home community by creating a unique relationship between the human and divine worlds; through Olympic victory Alkibiades demonstrated to all that he had the favor of the gods.[2] And, in recognition of Alkibiades' heroic achievement, and in gratitude for his proving Athens' own pan-Hellenic preeminence, the Athenians granted him free meals for life. At Athens all chariot victors at the crown games were admitted into a select "winner's circle" that feasted in the Prytaneion, at state expense.[3] In addition, the sense of accomplishment

[1] For the general discussion of ritual and display as social dialogue see C. Geertz, *The Interpretation of Cultures* (New York: Basic Books, 1973), *Local Knowledge: Further Essays in Interpretive Anthropology* (New York: Basic Books,1983), and J.J. MacAldoon, *Rite, Drama, Festival, Spectacle. Rehearsals Toward a Theory of Cultural Performance* (Philadelphia: University of Pennsylvania Press,1984). For the importance of ancient Greek athletic display see L. Kurke, *The traffic in praise : Pindar and the poetics of social economy* (Ithaca, NY: Cornell University Press, 1991); and M. Golden, *Sport and Society in Ancient Greece* (Cambridge: Cambridge University Press, 1998). For a more focused discussion of display at Athens see D. Kyle, *Athletics in Ancient Athens* (Leiden: E. J. Brill, 1987); P. Nichols, *Aristophanes' Novel Forms: The Political Role of Drama* (London: Minerva,1998); and S. Goldhill and R. Osborne, edd., *Performance culture and Athenian democracy* (Cambridge: Cambridge University Press, 1999).

[2] L. Kurke, "The Economy of *Kudos*," in L. Kurke and C. Dougherty, edd., *Cultural Poetics in Archaic Greece* (Oxford: Oxford University Press, 1988) 131-163.

[3] Along with the descendants of the tyrant-slayers Harmodios and Aristogeiton, "victors in the races for horses and chariots at the Olympic, Pythian, Nemean, and other games" were allowed to dine at state expense in the Prytaneion. *IG* I[3] 131.15-17.

generated by the Olympic victories helped to convince the Athenians that Alkibiades was also worthy of social and political responsibility, in short, that he was worthy of the Sicilian command which he sought.

The earliest literary sources, the Homeric epics, document some of the social motives behind elite animal production. For those who could raise sufficiently large herds, animal husbandry offered a means to accrue power and influence through the all-important ritual of feast-giving.[4] Through reciprocal gifts of food, feasting solidified friendship-obligation relationships among the semi-autonomous "Homeric" households, and even among the dependents and masters of these households themselves.[5] Homer often speaks of the importance of feasts in creating and maintaining a hero's reputation. In *Odyssey* 2.48, for example, Telemachos laments the fact that his mother's suitors are destroying his source of livelihood, feasting in Odysseus' house, slaughtering his cattle, sheep and goats. The slaughter is so upsetting to Telemachos because these animals represent the raw materials that he needs to demonstrate his aristocratic status to Homeric society. For the same reasons, Odysseus' household is attractive to the suitors, since whoever could win the hand of Penelope would also gain Odysseus' animals and lands.

In a similar manner, funeral games, where wealthy men pitted their chariot teams against those of their peers in return for prizes and honor, offered a forum in which the Homeric elite might renegotiate their status relationships and obligations.[6] The funeral games that

[4] M. Maus, "Essai sur le don. Forme et raison de l'échange dans les sociétés archaïques" *l'Année Sociologique*, seconde série, 1923-1924, was one of the first modern scholars to identify the significance of gifts in social networks. For Homeric gift relationships see B. Qviller, "The Dynamics of the Homeric Society," *Symbolae Osloenses* 56 (1981): 109-55; I. Morris, "The use and abuse of Homer,"*Class Ant.* 5 (1986): 81-138; and S. Hodkinson, "Imperialist Democracy and Market-Oriented Pastoral Production in Classical Athens," *Anthropolozoologica* 16 (1992): 53-61.

[5] See W. Donlan, "Reciprocities in Homer," *The Classical World* 75 (1982): 137-75.

[6] M. B. Poliakoff, *Combat Sports in the Ancient World* (New Haven: Yale University Press, 1987); I. Morris, *Death-Ritual and Social Structure in*

Achilles held for Patroklos best illustrate this system. For the Homeric heroes, victory at the games brought honor, proof of ability and divine support, while defeat brought ignominy and shame. Nestor, for example, warns his son Antilochos to ride close enough at the turns that he gains on his opponents, but not so close that he crashes his chariot, for if he crashes he will be a joy to his rivals and a disgrace to himself.[7] At Patroklos' games the heroes do compete viciously: Antilochos and Menelaos nearly wreck each other at one of the turns, and at the finish line a last-place Menelaos challenges Antilochos, accusing him of cheating, of purposely trying to wreck him. In the end, only Achilles' skillful arbitration prevents bloodshed. In some sense, then, these public horse and chariot contests must have served not only as fora where men could test their abilities (and their stables) against their peers without resorting to violence, but even as conflict resolution devices, where tensions between rivals might be constructively diffused.

The central role of animals within these negotiations of status did not end with the heroic age, as the loosely organized Homeric households were transformed into the more structured society of the polis. Although the focus for social and political authority began to shift from individual to community, the display and consumption of animals retained a central role in the expression of personal ambition and status. The feast for the household transformed into the feast for the community; what had been primarily a personal demonstration of loyalty, support and friendship between a warlord and his retainers and allies became a public celebration of the community, a confirmation of social roles and obligations between elected officials and their constituents. Likewise, the funeral races

---

Classical Antiquity (Cambridge: Cambridge University Press, 1992); and D. G. Kyle, *Sport and Spectacle in the Ancient World, Ancient Cultures* (Malden, MA: Blackwell, 2007).

[7] *Il.* 23.342-3. Nestor obviously believes that his youthful victories continue to exalt his status. Otherwise he would not bring them up before the games, nor would Achilles award him with a prize

and competitions transformed into the pan-Hellenic Crown Games (the Olympic, Pythian, Isthmian, and Nemean festivals), where elites continued to sort out not only their own personal status questions, but even the reputations of their respective communities as a whole.[8]

As many scholars have shown, the development of the polis was the result of a long-term balancing act between the interests of the emerging citizenry and those of the ruling great families, with only a gradual transfer of authority and responsibility away from the nobility to formal state institutions.[9] At Athens, for example, not until the reforms of Solon, and the institution of the state festival of the Genesia, did funerary celebrations and contests begin to separate from the exclusive control of the great families.[10] Likewise, it was not until the expansion of the Panathenaic festivals under Peisistratos and his sons, and the creation of state celebrations and cults by Peisistratos and Kleisthenes, that athletic and equestrian competition were given a more civic focus, and the primacy of the clan festivals was challenged by regional and local civic celebrations.[11]

This redirection of at least a substantial part of elite avenues for display was an important step in the incorporation of individual and family identity into the identity of the community. Such a gradual accretion of

[8] C. Morgan, "The origins of Panhellenism," in N. Marinatos and R. Hägg, edd., Greek Sanctuaries. New Approaches (London: Routledge, 1993), 26-27; and I. Morris, "The Art of Citizenship," in S. Langdon, ed., New Light on a Dark Age (Colombia, MO: University of Missouri Press, 1997), 9-43.

[9] E.g. R. Sealey, A History of the Greek City States 700-338 B.C. (Berkeley and Los Angeles: University of California Press, 1976); S. Humphreys, Anthropology and the Greeks (London: Routledge, 1978); A. Snodgrass, Archaic Greece: The Age of Experiment (Berkeley and Los Angeles: University of California Press, 1980); I. Morris, Burial and Ancient Society. The Rise of the Greek City-State (Cambridge: Cambridge University Press,1987); and J. M. Hall, A history of the archaic Greek world, ca. 1200-479 BCE (Malden, MA : Blackwell, 2007).

[10] R. Parker, Athenian Religion: A History (Oxford: Clarendon Press, 1996), 48-49.

[11] Kyle, Athletics in Ancient Athens and G. Anderson, The Athenian experiment: building an imagined political community in ancient Attica, 508-490 B.C. (Ann Arbor: University of Michigan Press, 2003).

functions to the community resulted in a progressive taming of the elite, as the emergent states attempted to make good citizens of their aristocrats by focusing their attention on civic interests, thus harnessing aristocratic resources to the service of the community. As a result, the notion of personal greatness in the Archaic and Classical periods comes increasingly to depend upon the idea of civic approbation. Aristocratic "gentlemen" could compete, but only in ways that benefited the community as a whole, through such public institutions as raising and racing horses, sponsoring sacrifices, funding choral contests, or outfitting warships. As Qviller observed:

> the competition for power and prestige among the Greek nobility, from about the seventh century onward, moved away from the display of wealth at home and the attraction of a personal following to displays of munificence in the city center and contests for political office and political support independent of personal ties.[12]

Although elite activity was never completely replaced by state investment, it was certainly re-directed to benefit the state. *Megaloprepeia*, or conspicuous consumption, served to advance not only individual interests but also civic concerns through such traditional pursuits as competitive athletics, equestrian events, and sacrificial displays. This system of community-oriented *megaloprepeia* grew out of the aristocratic gift exchanges of the Homeric heroes, especially the sacrifices, acts of hospitality such as the public feasts, and the keeping of horses.[13] In many respects *megaloprepeia* is the civic appropriation and circumscription of aristocratic competitive expenditure, the transformation of the private gift exchange into the public adornment of the city. To put it another way, at some point during the Archaic period, the nature of this lavish, personal expenditure changed

---

[12] Qviller, "The Dynamics of the Homeric Society," 120.

[13] Kurke, *Traffic in Praise*, 168-9. See also W. Burkert, *Homo Necans: The Anthropology of Ancient Greek Sacrificial Ritual and Myth* (Berkeley and Los Angeles: University of California Press, 1983).

from the private gifts of aristocratic gentlemen, employed to create bonds between aristocratic gentlemen, to very public, conspicuous displays, meant to emphasize the links between the man of wealth, his community, and the gods. As Aristotle says: "The *megaloprepes* spends not for himself but for the community."[14]

A selection from Xenophon's *Oeconomicus* best illustrates the complex relationships created and maintained by *megaloprepeia*:

> I see that it is necessary for you to offer many and great sacrifices, or I think that neither gods nor men would put up with you. Then it is your duty to receive many foreigners as guests, and those lavishly. Then also you must feast and benefit the citizens, or be bereft of allies. Still, in addition to that, I perceive that the polis commands you to spend a great deal of money in some ways already, on the raising of horses and the production of choruses and superintending the *palaistra* and accepting presidencies. But in the event of war, I know that they will demand of you trierarchies and property taxes so high that you will not easily bear them. And wherever you seem to do any of these things inadequately, I know that the Athenians will punish you no less than if they had caught you stealing their own property.[15]

Although Xenophon's Sokrates dwells on the element of compulsion in these duties, he does so simply to bolster his claim that Kritoboulos' wealth makes him pitiable, and thus a useful recipient for Sokrates' advice on the "proper" forms of estate management and public behavior. Apart from this rhetoric, however, the obligations of a rich man at Athens are clear: he should sacrifice often and abundantly, feast citizens and foreigners alike, raise horses, produce choruses, support a warship and even pay additional taxes in time of war.[16]

---

[14] *Nichomachian Ethics* 1123a.4-5.

[15] Xen. *Oec.*, 2.5-7.

[16] For *megaloprepeia* and liturgies see N. Lewis, "*Leitourgia* and related terms," *GRBS* 3 (1960): 175-184; A.W.H. Adkins, *Merit and Responsibility: A*

Yet the *megaloprepes* was not an altruist, and he did not act simply for the good of the community. As J. K. Davies has suggested, *megaloprepeia* among the Greeks represented "a deliberate investment in the goodwill of public opinion within the deme, tribe, or state."[17] A system of elite public display arose in the early sixth century to replace the traditional power-base of the aristocrats, the personal charisma based on noble birth and control over family cult and tribe. At Athens, for example, the sources present a picture of a sustained legislative onslaught on the hereditary powers exercised by the elite in Athenian cults and Athenian public life.[18] With the decline of cult-power, the Athenian gentleman came to rely on public magnificence and display to provide a new channel to political power. As a result, the wealthy expended their resources on lavish "gifts" to the community in order to gain that community's goodwill. This goodwill, *charis* in Greek, "was nothing less than the primary basis both of election to office and of preponderant political influence."[19] At the end of the fifth century, however, the goal of the

---

study in Greek Values (Oxford: Oxford University Press, 1960), 202-205, 211-214; *Moral Vales and Political Behaviour in Ancient Greece* (New York: W. W. Norton, 1972), 121-125; W. R. Connor, *The New Politicians of Fifth-Century Athens* (Princeton: Princeton University Press, 1971), 18-22; M. I. Finley, *Politics in the Ancient World* (Cambridge: Cambridge University Press, 1983), 24-49; *The Ancient Economy* (Berkeley and Los Angeles: University of California Press,1985), 150-154; and J. Ober, *Mass and Elite in Democratic Athens* (Princeton: Princeton University Press, 1989), 199-247.

[17] J. K. Davies, *Wealth and the Power of Wealth* (Salem, NH: Arno, 1981), 92.

[18] Cf. Lysias 30 where the speaker attacks one Nichomachos, who has been commissioned by the *demos* to revise the official sacrificial calendar. According to this speech, Nichomachos has removed 3 talents' worth of traditional, gene-controlled sacrifices. See Parker, *Athenian Religion*, 43-45, 218ff., for further discussion.

[19] Davies, *Wealth and the Power of Wealth*, 96. The connections between conspicuous public consumption and political power is best expressed by a speech from Lysias, written in 388/7 BCE: "There are, indeed, persons who spend money in advance, not with that sole object, but to obtain a return of twice the amount from the appointments which you consider them to have earned" (19.57). As Lysias observes, would-be candidates spend their money on public works, *megaloprepeia*, which in some way seems to have "earned" them the offices they desired.

megaloprepes shifted from political to forensic power. The Athenian elite now expended their resources in lavish displays not to gain votes at the polls but to gain votes in the lawcourts. A speech from Lysias best illustrates the new goals:

> I fitted out a trireme five times, and I fought naval battles four times, and I paid many special levies in war, and all the rest of the liturgies I performed, second to none of the citizens. And on account of this I spent more than what the city ordered, in order to be better thought of by you and, if somehow some misfortune should befall me, in order to contest it better.[20]

Essentially, the speaker is asking the jurors not to consider the charges and evidence against him but rather the good deeds he has done on behalf of the state.[21]

Although scholars such as J. K. Davies, D. Whitehead, and J. Ober have documented in some detail the general mechanics of using wealth to gain and maintain political and forensic power at Athens, no study has yet considered the wider social role that the conspicuous consumption of animals played in maintaining, and even creating, elite status and identity.[22] In actuality, the public gratitude and the power conspicuous display generated were neither purely Athenian nor purely political. The talismanic power acquired by victors at the pan-Hellenic games, for example, was both vaguer and more pervasive than Davies had realized, conveying much more than simply "gratitude at the polls." Indeed, certain types of conspicuous consumption such as a horse and chariot victory brought life-long honor and prestige both publicly, in the form of

---

[20] Lys. 25.12-13.

[21] Examples of these appeals are numerous: Andokides 1.149; [Andokides] 4.42; Lysias 3.47; 6.46; 7.30-31; 19.56-7; 21.12, 25; Isokrates 7.53; 16.35; Isaios 4.27-31; 5.35-38; 6.60-61; 7. 37-42; Demosthenes 20.151; 21.153; 25.76-78; 36.40-42; 38.25; 47.48.

[22] J. K. Davies, *Weath and the Power of Wealth*; D. Whitehead, "Competitive Outlay and Community Profit: Filotimia in Classical Athens," *Classica et Mediaevalia* 34 (1983): 55-74; and J. Ober, *Mass and Elite in Democratic Athens*.

state recognition and reward, and privately, through personal deference and respect.

Horse Displays

The horse was quite rightly regarded as an obvious sign of wealth and nobility in the ancient world. Aristotle, for example, believed that the number of horses a man owned indicated the degree of his wealth because "horse-breeding requires the ownership of great resources."[23] His sentiment is echoed by Xenophon, who claimed that keeping horses marked a man out as wealthy.[24] At Athens in particular, owning a horse, much more than any particular style of dress, differentiated the rich from the poor.[25] The reasons for this are clear: horses are expensive to maintain and unlike other large, expensive domestic animals, horses produce no marketable products such as meat, milk, or wool. Instead, they consume a great deal of resources in the form of grain, hay, training and equipment. In fact, horses were so expensive to maintain that the Athenian state during the Classical period was compelled to provide a cash supplement to each cavalryman, for each horse, in order to ensure the fitness (and continued existence) of its cavalry.[26] All of these financial drains prompted Isokrates to declare that the breeding of horses is possible only for gentlemen, only for those most blessed by fortune, something never to be pursued by one of low esteem.[27]

The prominence of the horse as a symbol of wealth was accentuated by the frequent public appearances of the official cavalry, whether as individuals or as a group; the Parthenon reliefs attest to the sheer impressiveness of horses moving in procession. As a matter of course, the mounted man would have offered a forceful reminder of his privileged status to his pedestrian fellows.

[23] *Politics* 1321a.5-15; cf. 1289b. 33-6.

[24] *Agesilaos* 9.6.

[25] Thucydides 1.6.5; Xenophon *Athenaion Politeia* 1.10.

[26] See I. G. Spence, *The Cavalry of Classical Greece* (Oxford: The Clarendon Press, 1993), Appendix 4.

[27] Isok. 16.23.

Demosthenes, for example, plays on the envy horses could arouse among those less fortunate when he critiques Meidias for "conveying his wife to the Mysteries and wherever else he wishes with a white chariot team from Sikyon."[28] Yet the Greeks did not keep horses primarily for their own private transportation; they spent their fortunes maintaining horses in an attempt to increase both their own and their state's reputations. The very ownership of expensive animals such as horses seems to have done much to create a man's reputation, to signal that he had "become" elite, become a gentleman. We have discussed earlier how, in the later fourth century BCE, Philip of Macedon gave several Athenian ambassadors a number of herds, flocks and horses in order to gain their support once they returned to Athens. This act roused the ire of Demosthenes, who contended that the Athenian people "instead of becoming angry and demanding the punishment of the traitors, stared at them, envied them, honored them, and considered them true men."[29] Philip's gift allowed these ambassadors to change their social status, to take part in elite activities such as chariot racing. It is this change in status, this envy and honor that the masses gave the ambassadors that upset Demosthenes so greatly.

The reason for Demosthenes' anger is that chariot racing was by far the most prestigious use to which horses could be put. Xenophon argues that the best activity of the *megaloprepes* is raising horses for competition in honor of the community.[30] Likewise, Demosthenes' contemporary Hypereides heaps praise on a certain Lykophron for continually overstretching his financial resources in attempts to breed better horses for the glory of the city.[31] Lysias explains what both the city and the horsebreeder

---

[28] 21.158. Cf. [Demosthenes] 42.24, where the orator chides Phainippos for adopting a chariot as his sole mode of transport.

[29] 19.265.

[30] *Oec.* 2.5-7.

[31] Hypereides 1.16. Xenophon (*Hiero* 11.5) speaks about the desirability of a city possessing the greatest number of horse-breeders and competitors at equestrian festivals.

hoped to gain through such acts of horse display: when Nikophemos was a cavalryman "he not only bought magnificent horses but also won horse races at the Isthmian and Nemean games so that the city was honored in public and he was crowned."[32] The city won public honor, Nikophemos the crown. According to Xenophon, that crown entitled the victor to public admiration and respect, in short the talismanic power about which Kurke has written.[33] A work attributed to Demosthenes explains nature of this talismanic power:

> [Chariot racing] in magnificence and majesty... resembles the power of the gods, moreover it provides the most pleasant spectacle comprised of the greatest number and largest variety of features, and is considered worthy of the greatest prizes. For, in addition to those prizes offered, exercising and practicing such skills in itself offers a worthy prize to those who even moderately desire manly excellence, arete.[34]

Thus, because of the splendor, scale and expense, chariot racing evokes the power of the gods and in so doing demonstrates the manly excellence (arete) of the horse-owner. In addition, chariot victories make it plain to all that the victor has the favor of the gods, and since he has been triumphant in this, so too will he be in other, more mundane affairs.

The connection between keeping horses and elite identity is especially clear in Herodotos. While discussing the origins of the Alkmaionidai, Herodotos asserts that the preeminence of this famous Athenian family derived from a single chariot victory at Olympia. The story goes that Kroisos, king of Lydia, gave the eponymous ancestor Alkmaion a sum of money so that he could purchase an excellent team of horses. After training and outfitting his team, Alkmaion went on to win the four-horse chariot race in the Olympic games of 592 BCE. This victory, Herodotos

[32] 19.63.
[33] Xen. Hiero 11.5. Kurke, "The Economy of Kudos," 131-163.
[34] [Dem.] 61.24-25.

argues, so ennobled Alkmeon and his entire family that they became established from that time forward as "illustrious."[35]

The close relationship between chariot-racing and political power is further exemplified by the career of Miltiades of Athens. Although Peisistratos held all power in Athens, Miltiades the son of Kypselos was also famous and influential, because his family kept four-horse chariots.[36] Clearly, Herodotos recognized that in Athens horse displays ennobled the displayer. From Pausanias we know the substance of Miltiades' displays: he won the Olympic four-horse chariot race in 560 BCE. As Herodotos says, this victory brought Miltiades social distinction and influence, but could bring him little political power so long as Peisistratos firmly controlled the political infrastructure at Athens. Since Miltiades had grand ambitions, he used his international reputation as an Olympic victor to raise an army and carve out a kingdom in the Thracian Chersonese. Once in power in the north, Miltiades continued to exhibit his status as a man of influence and authority through his Olympic victory. For many years he issued coins commemorating his Olympic triumph.[37] At some point, in order to further stress his Olympic connections, Miltiades even made a dedication at Olympia, a ivory horn set up in the treasury of the Sikyonians, the oldest dedication by a chariot victor at Olympia.[38] Clearly, Miltiades (and his supporters) believed that his chariot victory retained some power and was thus worthy of continued emphasis.

Miltiades' half-brother Kimon also built a reputation around Olympic chariot racing, and although he remained in Athens and was not so illustrious as his half-brother,

---

[35] 6.125.1.

[36] Hdt. 6.35.1.

[37] See Kyle, *Athletics in Ancient Athens*, 158.

[38] Pausanias 6.10.8; 6.19.6.

Kimon may have been more politically adept.[39] Notice how Herodotos summarizes his career:

> It befell his father, Kimon the son of Stesagoras, to flee Peisistratos the son of Hippokrates as an exile from Athens. And it happened that he won an Olympic chariot victory while in exile (and by this victory he carried off the same honor as his half-brother Miltiades). But after this, when he won his next Olympiad with the same mares, he handed it over to Peisistratos to be heralded as victor, and by handing the victory over to this man, he returned home on his own terms. And it befell him, when he won another Olympic victory with the same horses to die at the hands of the sons of Peisistratos.[40]

Kimon, exiled from home by the tyrant Peisistratos, used his Olympic victory to bargain for his return. By having Peisistratos' name announced as the true victor, Kimon both humbled himself and exalted the tyrant, who had never won such distinction with his own animals.[41] Such generosity and humility paved the way for Kimon's return home and probable reward, for Herodotos says that Kimon returned home "on his own terms." It is possible that Kimon gave up his victory to show Peisistratos that he was no rival, that he refused to enter politics, but the situation may be more complicated. Kimon certainly got the tyrant's attention through his Olympic victory, but we have no evidence that he refused to enter politics. In all likelihood, by giving up his god-given victory he proved to Peisistratos his loyalty, and ensured himself a place in the tyrant's government, whatever that may have entailed. After the Peisistratos' death, Kimon may have sought Olympic victory as the base for his own bid for the

---

[39] Compare the case of Pronapes, chariot victor and hipparch who prosecuted Themistokles c. 470. See Kyle, *Athletics in Ancient Athens*, 161, n. 31.

[40] 6.103.1-3.

[41] See Kurke, *Traffic in Praise*, 180, for other examples in which men "surrendered" their victories. The Sicilian tyrants in particular were famous for acquiring other men's achievements.

tyranny, an action that was perceived as a threat by Peisistratos' sons and resulted in Kimon's death.

Demaratos, the Eurypontid king of Sparta seems to have used an Olympic victory in much the same way as Kimon, to gain political support at home. Herodotos tells us that Demaratos won an Olympic victory (in 508 or 504 BCE)[42] which he then "handed over to the state, the only Spartan king to do so."[43] In effect, by dedicating his victory to Sparta, as Kimon had done for Peisistratos, Demaratos was not only bargaining for the Spartans' support but also publicizing his generosity and heroic ability. Demaratos was driven to such an act because throughout this reign (c. 515-491 BCE) he and his Agiad colleague Kleomenes were engaged in a bitter struggle for control of Spartan foreign policy. The main climaxes of their struggle are well known: c. 506 Demaratos prevented Kleomenes from invading Attika and deposing the fledgling Athenian democracy, and in 491 he prohibited Kleomenes from removing a pro-Persian faction on Aigina.[44] Unfortunately, it was this support of pro-Persian factions that gave Kleomenes the local as well as pan-Hellenic support required to banish Demaratos from Sparta, on a false charge of illegitimacy. The role of Demaratos' Olympic victory in the struggle with Kleomenes is clear: by dedicating his victory to the Spartans, Demaratos was able to gather the authority necessary to recall Kleomenes from Athens. By proving his worthiness to the gods at Olympia he was able to prove the strength and "rightness" of his policy to the Spartans.

After Demaratos, chariot victories seem to take on a special resonance for the Spartans. Pausanias claims that "after the Persian invasion the Lacedaemonians were the

---

[42] A. Hönle, *Olympia in der Politik der griechischen Staatenwelt* (Bebenhausen: Lothar Rotsch, 1972), 129, 150ff., who follows Moretti in tentatively dating the victory to 504 BCE, admits that this date is by no means secure.

[43] 6.70.

[44] Hdt. 5. 75ff.; 6.65. See P. Cartledge, *Sparta and Lakonia. A regional history 1300-362 BC* (London: Routledge, 1979), 199ff, for a discussion of these events.

most enthusiastic of all Greeks in hippotrophia."[45] In all likelihood, the Spartans used the crown festivals and other regional games to assert and maintain their superiority over the other Peloponnesians, to show that the gods sanctioned their actions and their rule. One Spartan citizen, a certain Damonon won more than thirty equestrian victories in total at Olympia and at regional Lakonian festivals held in both Spartan and Perioikic territory: the Potidaia at Sparta, Helos and Thouria; the Athenaia and Ariontia at Sparta; the Eleusinia at modern Kalyvia; the Lithesia near Cape Malea; and the Parparonia in the Thyreatis, held at the site of the great victory over the Argives in *c.* 550 BCE.[46] The victory of a Spartan at this controversial site on the border of Argos must have done much to maintain notions of Spartan superiority and divine favor.

By the late fifth century, the connection between elite reputation and horses was so axiomatic at places like Athens that the comic playwright Aristophanes lampooned it. In the *Clouds*, Pheidippides, a young would-be aristocrat, was so obsessed with horses that he dreamed of them at night and squandered the family fortune in an effort to acquire a winning team of racers.[47] But even the exaggerations of comedy could not anticipate the grandest of all Athenian chariot displays, the Olympic victories of Alkibiades. In 415 BCE, Alkibiades entered seven four-horse chariots in the Olympic games. As a result, in the same event, he won first, second and fourth. Alkibiades took on this grand expense for much the same reasons as earlier Athenians: to acquire personal glory and to bring honor to Athens. According to Thucydides, Alkibiades raced his seven chariots in an effort to convince the other Greeks that Athens was not weak, as all had supposed, but even greater and more powerful than anyone had realized.[48] In return for increasing the prestige of Athens,

[45] 6.2.1-2.
[46] See Cartledge, *Sparta and Lakonia*, 233.
[47] *Nubes* 21-22.
[48] Thuc. 6.16.1-4.

however, Thucydides has Alkibiades argue that he has a prescriptive right to ask a favor—election to commander of the upcoming Sicilian expedition. The Athenians clearly agreed, even though Alkibiades' rival for the Sicilian command, the elder statesman Nikias, spoke at length about Alkibiades' lack of military experience and urged the demos not to grant him the command simply because of his expenditure on horse-racing.[49] In the end, Nikias was unable to sway popular opinion, and Alkibiades was elected to the command because he was a proven winner, a man with the demonstrated support of the gods. As Kurke has shown, victory at the crown games was linked in the Greek mind with victory in battle.[50] After all, victory in any sense was a god-given quality. Indeed, Alkibiades' chariot victories at Olympia overshadowed his later military blunders, for despite his subsequent failure in the Sicilian command and his later betrayal of Athens, almost a century later Demosthenes still praised Alkibiades' Olympic chariot victories and counted them on the credit side of his checkered career.[51]

After Alkibiades' grand claims and magnificent displays in 415 BCE, horse displays at Athens seems to have declined. Davies has argued that this was a conscious decision by the elite to focus their attentions on other forms of conspicuous public display.[52] Kurke agrees and suggests that the excesses and failures of Alkibiades convinced the public that there was little merit in granting high office to inexperienced young men, whose only claim to fame was hippotrophy.[53] Although an undercurrent of dissatisfaction is clearly recognizable in the ancient literary sources, this may be mere rhetoric from those new politicians such as Kleon whose families had no tradition of hippotrophy or could not afford to build one. I. G. Spence offers a better explanation for the apparent Athenian lack

[49] See Thuc. 6.12 for Nikias' speech.
[50] Kurke, *Traffic in Praise*.
[51] Dem. 21.144-5.
[52] Davies, *Wealth and the Power of Wealth*, 95.
[53] Kurke, *Traffic in Praise*, 1991: 177ff.

of interest in horse breeding by suggesting that the decline stems from the participation of the cavalry in the oligarchic coup of 404 BCE.[54] As [Lysias] 20.13 asserts, the cavalry were among the most active supporters of the Thirty, even in the final stages. In fact, the cavalry conducted military campaigns against the democrats at Phyle and Peiraios and was responsible for the execution of all the male citizens in Eleusis so the Thirty could use it as a base of operations.[55] According to Spence, even though the Athenians quickly forgave the cavalry, they were slow to forget the betrayal. Financial sanctions were imposed upon those who had personally served under the Thirty, and for some years service in the cavalry was seen as a cowardly act, an avoidance of civic responsibility.[56] For many, keeping horses remained a cause for suspicion, a sign of questionable loyalty, and all equestrian commemoration, both official and private, lapses. The Athenian cavalry class was not fully rehabilitated until their spectacular victory at Mantineia in 362 BCE. After this heroic act the stigma of cowardice recedes from cavalry service, and private individuals once again begin to erect equestrian funerary monuments. Moreover, Athenians once again became involved in public-minded hippotrophy as an elite-bashing statement from Lykourgos illustrates:

> Horse breeding, a munificent choregia, and other expensive gestures, do not entitle a man to any such recognition from you, since for these acts he alone is crowned, conferring no benefit on others. To earn your *charis* he must, instead, have been distinguished as a trierarch, or built walls to protect his city, or subscribed generously from his own property for the public safety. These are services to the state: they affect the welfare of you all and prove the loyalty of the donors, while the others are evidence of nothing but the wealth of those who have spent the money.[57]

[54] Spence, *The Cavalry of Classical Greece*, 210ff.
[55] Xen.*Hell.* 2.4.8-10.
[56] Lys. 16.13ff.
[57] Lykourgos *Leokr.* 139-140.

For the speaker to place horse breeding in first position among the excesses of the elite and to attack it as an unsuitable way to earn public goodwill shows that horse displays played an active role in elite status negotiations during the latter half of the fourth century.

For Greeks outside Athens, however, participation at the crown games continued unabated after the Peloponnesian Wars. In 396 and 392 BCE, Kyniska, daughter of Archidamos and sister of Spartan kings Agis and Agesilaos, entered and won the four-horse chariot race at the Olympia.[58] With these two victories she became the first woman ever to win at a crown game, bringing herself, her family and her community everlasting glory. Plutarch says that she was encouraged in her equestrian pursuits by her brother King Agesilaos. It is likely that Kyniska's chariot victories were a bid for popular support by Agesilaos who was then engaged in a power struggle with the ephorate over Sparta's role in the wider Greek world.[59] Agesilaos wanted Sparta to take a more assertive role in ruling Greece, but the ephors were adamant that he mind matters closer to home. In fact, Agesilaos had already been fined by the ephorate for garnering support among the Spartiates, as Plutarch put it he was fined for "acquiring the common citizens as his own."[60] Perhaps Kyniska's victories were a sly way for Agesilaos to solicit popular support without personally rousing the ire of the ephors. In any event, the statement made by Kyniska's victories is clear: the gods supported the Eurypontid house.

Another regal individual who made much of victory at the crown games was Philip of Macedon. In 356 BCE, Philip won the two-horse chariot race at Olympia.[61] Immediately, he began to issue large gold and bronze coins depicting his chariot team to celebrate and publicize the

---

[58] Paus. 3.8.1; Plut. *Agesilaos* 20.1; *Inschriften von Olympia*, 160.
[59] See Hönle, *Olympia in der Politik der griechischen Staatenwelt*, 146-59.
[60] *Agesilaos* 5.
[61] Plutarch *Alexander* 4.9; and *Moralia* 105a.

event.[62] In this action he recalls Miltiades, another Olympic victor who ruled in the north of Greece. In 350, when Philip seized the neighboring kingdom of Epiros, he suspended the local Molossian coinage and replaced it with that commemorating his Olympic triumph.[63] Seemingly, Philip sought to commemorate one victory with another; perhaps he stressed the chariot victory to show his new subjects that he had the divine support and heroic ability to rule. Perhaps he used it to assert his common Greek identity and heritage. In a similar fashion, upon his defeat of the Chalkidian League in 348, Philip re-issued his Olympic coinage. Philip clearly wished to remind the rest of Greece that his military victories were sanctioned by the gods.

Sacrifice

Just like victory in the crown games, wealth in healthy sacrificial animals was also seen as a sign of divine favor. From Homeric down through Classical times men attributed the health and prosperity of their herds and flocks to the goodwill of the gods. In the *Iliad* the poet observes that a certain Phorbas became rich in flocks through the grace of Hermes, the god of flocks.[64] Likewise, the Homeric Hymn to Earth, says that "the man whom Earth best favors has fields abounding in herds, and both great material prosperity and wealth."[65] Some time later, Hesiod asserted that only the noblest men, those who have been blessed by the gods, are rich in flocks (WD 307). And in the fifth century, Aischylos has the *Eumenides* bless the Athenians with herds and crops.[66] Since wealth in sacrificial animals is therefore a mark of divine support, the public display of such animals at a sacrificial dedication makes plain to all spectators both the depth of a man's god-

---

[62] N.G.L. Hammond, *Philip of Macedon* (Baltimore: Johns Hopkins University Press, 1994), 113ff.

[63] Hammond, *Philip of Macedon*, 51, 120.

[64] *Il.* 14.490.

[65] HH 30.10.

[66] Aischylos *Eumenides* 939-44.

favoredness and his generosity. Besides, it is always good to remind the divine of one's piety, humility and thankfulness.

The literary evidence for such sacrificial dedications by private individuals is rare for the Archaic and early Classical periods. One of the earliest and most dramatic dedications is that of Kroisos, King of Lydia, in the early sixth century. Upon receiving a favorable oracle from Delphi, the answer to his test question about what he was doing on a certain day, Kroisos dedicated 3000 beasts to Delphian Apollo and commanded all of his citizens to sacrifice to the god whatever they could afford.[67] Kroisos did this, so Herodotos says, in order to win the favor of the god. He certainly achieved his goal. In return for the animals and other gifts of gold and silver, the Delphians gave Kroisos and all his Lydians the right to consult the oracle before all others (Greeks included), the chief seats at the festivals, and the perpetual right of Delphic citizenship for any Lydian who wished it.[68] The fact that the Delphians granted such honors to a barbarian king, show the power of sacrificial dedications. Kroisos' generosity greatly impressed upon the Delphians his piety, his resources, and above all his general preeminence.

The only other large-scale sacrificial dedication reported by the literary sources is that of Jason, tyrant of Pherai. As we discussed in Chapter I, in the year 370 BCE, Jason, tyrant of Pherai and warlord of all Thessaly, ordered his subjects to gather one thousand cattle, along with over ten thousand sheep, goats and swine for an offering to Apollo at the upcoming pan-Hellenic Pythian Games. According to Xenophon, Jason intended by this spectacular dedication to become director not only of the festal assembly in honor of the god but even of the Pythian Games which followed. Both were highly visible and symbolic positions in their own right, and not without a certain measure of authority, as the director of the Games presided over the various

[67] Hdt. 1.50.
[68] Hdt. 1.54.

competitive events. Although Jason died before he or his
parade could leave Pherai, the fact that he organized this
display with the intent of attaining a prestigious and highly
visible office suggests that animal displays, especially large,
conspicuous ones, held a certain symbolic resonance in
contemporary Greek society.

The remainder of our information about private
sacrifice comes from inscriptions, largely from fourth-
century Athens. Unfortunately, the Athenian bias of the
evidence poses several methodological problems because in
democratic Athens the demos came to replace the wealthy
individual as the leading patron and provider of animals for
sacrifice. As a result, the Athenian elite tended to produce
animals for sale at the public sacrificial market, instead of
dedicating them in elaborate sacrifices such as Jason's. The
money these sales generated allowed the Athenians to
engage in other prestige displays such as outfitting a
warship and sponsoring a chorus. Thus, animal production
at Athens aided in maintaining an individual's social
position tangentially, by providing him the marketable
resources with which to maintain his other liturgical
obligations. Still, animal production at Athens does seem to
have an active liturgical role. The numerous tribal and
deme sacrifices allowed ample opportunity for the wealthy
man to enhance his standing among his neighbors by
donating sacrificial victims directly to public sacrifices, or
even allowing an individual to supply lavish private
sacrifices, at which all participants might have a substantial
share in the meat.

Even though most sacrifices on the polis-level were
provided by the state, paid from state funds, the sacrifices
conducted on a more local level, by the deme or phratry,
were not so adequately funded or so carefully controlled.
Consequently, officials of these smaller units might
supplement or even provide all the victims needed from
their own resources. While these contributions could be in
the form of money for purchasing victims, it would be
much easier for the local Athenian officials simply to

contribute the victims themselves, from their own herds.[69] In this way, no money had to change hands and the contributor still received the desired public acclaim by expending only a few of his excess sheep, as individual sacrificial events rarely required more than three or four sheep or goats. It is clear from the wording of the inscriptions that the men who sponsored these local sacrifices and feasts expected to get something in return for their efforts.[70] The inscriptions all say that the honorees acted because of *philotimia*, "love of honor," and this competition for public honor is yet another example of the state curbing private competition and redirecting it into the service of the community.[71] Whitehead captured this dynamic best: "As far as the polis was concerned, *philotimia*, even an excess of it, could be not merely accommodated but actively welcomed, provided always that the community itself was acknowledged to be the only proper source of *time* [honor] and thus the only proper object of the energy and (in particular) the expense that the *philotimos* sought to lay out."[72] As with earlier forms, this fourth-century expression of *philotimia* played a major part in the active build-up of a political goodwill that might be exploited as a lever to office and as a refuge in times of trouble.

Such exploitation of public goodwill was certainly the goal of a social climber in the Attic deme of Lamptrai. At some point in the latter half of the fourth century, the

[69] Unlike major, regional sanctuaries like Delphi and Eleusis, local cults did not have the resources to raise sacrificial animals to sell for sacrifice, so suppliants would have to bring their own or by them from private producers.

[70] Even though the bulk of the evidence comes from the last third of the fourth century, the lack of similar documentation from earlier times may be nothing more than an accident of loss and survival. Clearly, the system known from fourth-century inscriptions was not new; sacrifice and private feasting as a means to gain support had a long history among the ancient Greeks, with origins stretching back at least to Homeric times.

[71] *IG* ii^2 1204 from the deme of Lamptrai; *IG* ii^2 1163 from the tribe of Hippothontis. See V. Rosivach, *The System of Public Sacrifice in Fourth-Century Athens* (American Classical Studies 34) (Atlanta: Scholars Press,1994) 130, n. 76, 77, for other examples.

[72] D. Whitehead, " Competitive Outlay and Community Profit," 59-60.

citizens of Lamptrai rewarded a newcomer from Acharnai for his generosity in providing sacrifices by voting him the right of *ateleia*, "freedom from public duties," and a full share of the sacrificial meats equal to that of the demesmen.[73] Essentially, this Acharnian used his animal resources to gain acceptance in his new deme, as a local man of some importance. Although it is unclear whether he provided the sacrifices from his own herds or paid for their purchase, it is certain that his sacrifice gained him a certain amount of official public recognition and good-will.

In addition to sponsoring sacrifices, and the meals of meat they produced, the wealthy might also hold public feasts in order to establish an "account" of public good-will. For example, a certain Euthydemos of Eleusis used the occasion of his term as demarch to feast the demesmen at his own expense and to receive in return the same right of *proedria*, "sitting in the front seats" of the local theatre that his ancestors had received in their day.[74] This exchange of favors, the feast for the public honor of the front seat, has a certain ritualistic flavor to it that suggests a long history. Perhaps Euthydemos' ancestors had feasted their fellows in much the same manner in order to gain the honor of "front seats."

***   ***   ***

The wealthy elite of ancient Greece required animals to demonstrate and maintain their gentlemanly social status. They achieved this through a complex system of conspicuous displays on behalf of the community. The pan-

---

[73] *IG* ii² 1204. Such grants of *ateleia* equivalent to the better attested Roman practice of granting *immunitas*, suggest that Greek *poleis* regularly rewarded citizens with tax and other public exemptions in return for community service. In practice, granting *ateleia* allowed communities to acquire much needed public works, since such official grants of "freedom" encouraged the recipient to continue to spend on the community's behalf, if only to retain his "recognized" elite status. See R. Duncan-Jones, *Structure and Scale in the Roman Economy* (Cambridge: Cambridge University Press,1990),161-162; 178-182.

[74] Cf. Demosthenes (21.156), who argues that he feasted his fellow tribesmen in addition to performing other public services, and Isaios 3.80, in which a man feasts the female citizens of the deme on behalf of his wife.

Hellenic horse and chariot contests, for example, offered arenas where elites might sort out not only their own personal status questions, but even the reputations of their respective communities as a whole, as men tested their fortunes against their peers without resorting to violence. In addition, because the crown games were religious festivals, victory in an event bestowed a certain talismanic power on both the victor and his home community, demonstrating to all that the victor had the unquestionable favor and support of the gods. Consequently, an equestrian victory at the great games brought life-long honor and prestige, in the form of international recognition and reputation. In a similar fashion, sacrificial dedication demonstrated to all spectators the preeminence of the dedicator as well as his public generosity, since the ancient Greeks viewed wealth in the form of healthy sacrificial victims as a particular sign of divine favor.

In practice, these acts of community-oriented consumption ranged from the extraordinary displays done by such men as Miltiades, Alkibiades and Jason to the more modest but still significant bequests and displays made by individuals earning public thanks for providing one or two sacrificial victims for a public event. Nevertheless, whatever the scale, *megaloprepeia* made it plain to all that the sponsor had received the favor of the gods, and as such was worthy of honor and respect. In the end, it is significant that Greeks from Alkmaion of Athens to Philip of Macedon consciously attributed their elite reputations to their acts of *megaloprepeia*.

AFTERWORD
PASTORAL POLITICS AND ANCIENT HISTORY

Despite a century of vigorous scholarship our knowledge of ancient Greek animal husbandry, the politics it inspired, and the patterns of land use that it shaped has remained woefully incomplete, scattered across many different (and highly technical) sub-fields and as such has remained isolated from mainstream Greek historical discussions and teaching. So what? Why should ancient historians care about a highly technical subfield? This book has attempted to suggest that patterns of animal husbandry matter because they help explain wider choices. Animals were important to the rural lives of Archaic and Classical Greeks, something we Moderns tend to overlook, unless we have family members involved in agricultural production. But apart from giving a better background to rural life, understanding animals and the choices their producers made is also useful for contextualizing a host of other issues that impact the lives of more people than just specialist producers, from politics to war, to sacrifice, to sacred land, to wealthy conspicuous consumption. Animals were important (and desirable) to Greek men and women at all levels of society because they could, by their very nature, represent tangible, renewable wealth. They were a naturally exclusive status symbol—large animals such as cattle and horses could make an impression of great wealth because the expense of their upkeep, in terms of land and human labor, usually prohibited all but the richest citizens from owning them. But that did not mean that other members of society didn't try, as attested by Aristophanes' character Pheidippides, who is plagued by dreams of being

able one day to raise horses, and the ambassadors to
Macedon that so work up Demosthenes.

By focusing on the demands and goals of elite animal
producers I have attempted to give a new perspective on
some old questions. The connections shown here between
gentlemanly animal production and wider community
concerns such as land use, regulation and armed protection
of resources, and public entertainment bring a complexity
and perspective that I hope will be useful for both the
scholarship and teaching of ancient history. In our modern
technologically dominated world we rarely think about
how physically impressive a one-ton bull is. We also never
consider the time, effort and resources required to turn a
newborn calf into a Grand Champion.

What I find especially significant is that despite all of
the social, political and even environmental changes in
Greek society from 800 to 300 BCE, horses, cattle, sheep,
goats and swine did not lose their power to impress.
Powerful individuals such as Homer's Odysseus,
Peisistratos and Alkibiades of Athens, Jason of Pherai and
Philip of Macedon still made much of the prestige they
derived by playing pastoral politics. And despite the fact
that the ways in which these animals were raised, the
specific "animal management strategies" involved,
depended upon regionally varying cultural and
environmental characteristics, Greeks from all regions of
Greece tried to raise them. As a result, animal management
strategies were very much products of a specific time and
place, but their social import, though waxing and waning
for unique cultural and social reasons, did not wholly lose
its force. Although animal husbandry in Greece was a
complicated affair, dependent on such variables as proper
pasture, adequate markets, local systems of land use, and
the individual social and economic concerns of the elite, we
should not get bogged down in that complexity. While
comparisons between geography, climate, and even social
values can and often should be made, the inherent
variability of the Greek environment and experience
prohibits any overarching formula such as agro-

pastoralism or transhumance from describing the nuances of local tradition and method. This work has proposed a new model that is sensitive to regional variation, that embraces a multiplicity of different management strategies by focusing on their social impact and socio-cultural roles. The way beyond specialization and fossilized models is wider inclusion into social, cultural and political historical debates. I hope that future research can sort out the many regional responses to animal production, rather than expend energy searching for one general model that explains the Greek "experience," and in so doing, draw ancient agriculture out from the narrow confines of subsistence economics into which it has all too often been relegated. I have every hope that the future of pastoral politics will be exciting.

BIBLIOGRAPHY

Adkins, A.W. H. *Moral Vales and Political Behaviour in Ancient Greece*. New York: W. W. Norton, 1972.

———. *Merit and Responsibility: A study in Greek Values*. Oxford: Oxford University Press, 1960.

Alcock, S. and R. Osborne. *Placing the Gods. Sanctuaries and Sacred Space in Ancient Greece*. Oxford: The Clarendon Press, 1994.

Anderson, G. *The Athenian experiment: building an imagined political community in ancient Attica, 508-490 B.C.* Ann Arbor: University of Michigan Press, 2003.

Barthélémy, J. J. *Voyage du jeune Anacharsis en Grèce*. Paris: Debure, 1788.

Berthiame, G. *Les rôles du mágeiros. Étude sur la boucherie, la cuisine et le sacrifice dans la Grèce ancienne*. Mnemosyne Suppl. 70. Leiden: Brill, 1982.

Bintliff, J. L. and A. M. Snodgrass. "The Cambridge-Bradford Boeotian Expedition: the first four years." *Journal of Field Archaeology* 12 (1985): 123-161.

Bottema, S. "Palynological Investigations in Greece with Special Reference to Pollen as an Indicator of Human Activity." *Palaeohistoria* (1982): 251-89.

———. "Palynological Investigations on Crete." *Review Palaeobotany and Palynology*, 31 (1980): 193-217.

———. "Pollen Analytical Investigations in Thessaly." *Palaeohistoria*, 21 (1979): 20-39.

Bommlejé, L. S. and P. K. Doorn, edd. *Stroúza Region Project (1981-1993): an historical-topographical fieldwork*. Utrecht, 1984.

Bowden, H. *Classical Athens and the Delphic Oracle. Divination and Democracy*. Cambridge: Cambridge University Press, 2005.

Brendel, O. *Die Schafzucht im alten Griechenland*. Diss., Giessen, 1934.

Buckler, J. *Philip II and the sacred War*. Leiden: E. J. Brill, 1989.

————. "Thebes, Delphoi, and the Outbreak of the Third Sacred War." In P. Roesch and G. Argoud, edd., *La Béotie antique*, 237-246. Paris: Éditions du Centre national de la recherché scientifique, 1985.

Bugh, G. R. *The Horsemen of Athens*. Princeton: Princeton University Press, 1988.

Burford, A. *Land and Labor in the Greek World*. Baltimore: Johns Hopkins University Press, 1993.

Burkert, W. *Homo Necans*. Berekeley and Los Angeles: University of California Press, 1983.

————. *Greek Religion*. Cambridge, MA: Harvard University Press, 1985.

Campbell, J. K. *Honour, Family, and Patronage: a study of institutions and moral values in a Greek mountain community*. Oxford: The Clarendon Press,1964.

Cartledge, P. "Raising hell? The Helot Mirage—a personal review." In N. Luraghi and S. E. Alcock, edd., *Helots and their masters in Laconia and Messenia: histories, ideologies, structures*, 12-30. Cambridge, MA: Harvard University Press, 2003.

————. *Money, labour, and land: approaches to the economies of ancient Greece*. London: Routledge, 2002.

————. *Sparta and Lakonia. A regional history 1300-362 BC*. London: Routledge, 1979.

Chandezon, C. *L'élevage en Grèce (fin Ve-fin Ier s. a.C.). L'apport des sources épigraphiques*. Bordeaux: Ausonius, 2003.

Chang, C. "Ethnoarchaeological survey of pastoral transhumance sites in the Grevena region, Greece." *Journal of Field Archaeology* 20 (1993): 249-264.

————. "Pastoral transhumance in the Southern Balkans as a social ideology: ethnoarchaeological research in northern Greece." *American Anthropolgist* 95 (1993): 687-703.

Chaniotis, A. *War in the Hellenistic World*. Malden, MA: Blackwell, 2005.

————. "Habgierige Götter, habgierige Städte. Heiligtumsbesitz und Gebietsanspruch in den kretischen Staatsverträgen." *Ktema* 13 (1988): 21-39.

Cherry, J. "Pastoralism and the Role of Animals in the Pre- and Protohistoric Economies of the Aegean." In C. R. Whittaker, ed., *Pastoral Economies of Ancient Greece and Rome.* (Cambridge Philological Society, Suppl. Vol. 41), 196-209. Cambridge: Cambridge University Press, 1988.

Connor, W. R. "Early Greek Land Warfare as Symbolic Expression." *Past and Present* 119 (1988): 3-29.

————. *The New Politicians of Fifth-Century Athens.* Princeton: Princeton University Press, 1971.

de Polignac, F. *La Naissance de la cité greque.* Paris: Edition La Découverte, 1985.

Dahl G. and A. Hjort. *Having Herds. Pastoral Herd Growth and Household Economy* (Stockholm Studies in Social Anthropolgy 2). Stockholm: University of Stockholm Press, 1976.

Daverio Rocchi, G. *Frontiera e Confini nella Grecia Antica.* Roma: <<L'Erma>> di Bretscheider, 1988.

Davies, J. K. *Athenian Propertied Families 600-300 B.C.* Oxford: The Clarendon Press, 1971.

————. *Wealth and the Power of Wealth.* Salem, NH: Arno, 1981.

Detienne, M. and J.-P. Vernant, edd. *The Cuisine of Sacrifice among the Greeks.* Chicago: University of Chicago Press, 1989.

Dever, W. G. *Who were the Early Israelites and Where did they come from?* Grand Rapids, MI: Eerdmans, 2003.

Dixon, M. "*IG* IV2.1.75+ and the Date of the Arbitration between Epidauros and Hermion." *ZPE* 137 (2001): 169-173.

Donlan, W. "The Relations of Power in the Pre-State and Early State Polities." In L. G. Mitchell and P. J. Rhodes, edd., *The Development of the Polis in Archaic Greece,* 39-48. London: Routledge, 1997.

————. "Reciprocities in Homer." *The Classical World* 75 (1982): 137-75.

————. *The Aristocratic Ideal in Ancient Greece.* Lawrence, Kansas: Coronado Press, 1980.

Dow, S. "The Greater Demarkhia of Erkhia." *BCH* 89 (1965): 180-213.

Duncan-Jones, R. *Structure and Scale in the Roman Economy.* Cambridge: Cambridge University Press, 1990.

Durkheim, E. *De la division du travail social.* Paris: F. Alcan, 1911.

Edwards, A. *Hesiod's Ascra.* Berkeley and Los Angeles: University of California Press, 2004.

Ellis, J. R. *Philip II and Macedonian Imperialism.* London: Thames and Hudson, 1976.

Ensminger, E. M. *Beef cattle science* (7th ed.). Danville, IL: Interstate Publishers, 1997.

————. *The stockman's handbook* (7th ed.). Danville, IL: Interstate Publishers, 1992.

Finley, M. I. *The Ancient Economy* (2nd ed.). Berkeley and Los Angeles: University of California Press, 1985.

————. *Politics in the Ancient World.* Cambridge: Cambridge University Press, 1983.

————. *The World of Odysseus.* London: Chatto and Windus, 1978.

————. "Homer and Mycenae: Property and Tenure." *Historia* 6 (1957): 133-59.

————. *Studies in Land and Credit in Ancient Athens 500-200 B. C.: the horos inscriptions.* New Brunswick, N.J: Rutgers University Press, 1952.

Fisher, N. and H. van Wees, edd. *Archaic Greece: New Approaches and New Evidence.* London: Duckworth, 1998.

Forbes, H. A. "The identification of pastoralist sites within the context of estate-based agriculture in ancient Greece: beyond the Transhumance versus Agro-pastoralism debate." *ABSA* 90 (1995): 325-338.

Foxhall, F. "Cargoes of the Heart's Desire. The character of trade in the archaic Mediterranean world." In N. Fisher and H van Wees, edd., *Archaic Greece: New Approaches and New Evidence,* 295-309. London: Duckworth, 1998.

————. "'Greek Agrariansim.' Review of Victor Davis Hanson, *The Other Greeks.*" *Classical Review* 48.2 (1998): 390-1.

————. "Bronze to Iron: Agricultural Systems and Political Structures in Late Bronze Age and Early Iron Age Greece." *ABSA* 90 (1995): 239-250.

Frazer, J. G. *Pausanias's Description of Greece.* London: Biblo and Tannen, 1898.

Gallant, T. W. *Risk and Survival in Ancient Greece. Reconstructing the Rural Domestic Economy.* Stanford: Stanford University Press, 1991.

Garnsey, P. *Food and Society in Classical Antiquity*. Cambridge: Cambridge University Press, 1999.

―――."Mountain Economies in Southern Europe." In C. R. Whittaker, ed., *Pastoral Economies of Ancient Greece and Rome* (Cambridge Philological Society, Suppl. Vol. 41), 196-209. Cambridge: Cambridge University Press, 1988.

―――. *Famine and the Food Supply in the Graeco-Roman World*. Cambridge: Cambridge University Press, 1988.

Geertz, C. *Local Knowledge: Further Essays in Interpretive Anthropology*. New York: Basic Books, 1983.

―――. *The Interpretation of Cultures*. New York: Basic Books, 1973.

Georgoudi, S. "Quelque problèmes de la transhumance dans la Grèce ancienne." *REG* 87 (1974): 155-85.

Gernet, L. "*L'Approvisionnement d'Athènes en blé au Ve et au IVe siècle*." Université de Paris, Bibl. Fac. Lett. 25. Paris: Mélanges d'histoire ancienne, 1909.

Golden, M. *Sport and Society in Ancient Greece*. Cambridge: Cambridge University Press, 1998.

Goldhill, S. and R. Osborne, edd. *Performance culture and Athenian democracy*. Cambridge: Cambridge University Press, 1999.

Grove, A. T. and O. Rackham. *The nature of Mediterranean Europe : an ecological history*. New Haven, CT and London: Yale University Press, 2001.

Hall, J. M. *A History of the Archaic Greek World ca. 1200-479*. Malden, MA: Blackwell, 2007.

Halstead, P. "Traditional and Ancient Rural Economy in Mediterranean Europe: Plus Ça Change?" *JHS* 107 (1987): 77-87.

―――. "Present to Past in the Pindhos: Diversification and Specialization in Mountain Economies." *Rivista di Studi Liguri* 56 (1990): 61-80.

―――. "Pastoralism or household herding? Problems of scale and specialization in early Greek animal husbandry." *World Archaeology* 28 (1996): 20-42.

Halstead, P. and J. O'Shea, edd. *Bad year economics : cultural responses to risk and uncertainty*. Cambridge: Cambridge University Press, 1989.

Hammond, N.G.L. *Philip of Macedon*. Baltimore: Johns Hopkins University Press, 1994.

———. "Diodorus' narrative of the sacred war." *JHS* 57 (1937): 44-77.

Hansen, M. H., ed. *The Polis as an Urban Centre and as a Political Community, Acts of the Copenhagen Polis Centre 4*. Copenhagen: KDVS, 1999.

———. *The shotgun method : the demography of the ancient Greek city-state culture*. Columbia, MO: University of Missouri Press, 2006.

Hansen, M. H. and T. H. Nielsen, edd. *An inventory of archaic and classical poleis*. Oxford: Oxford University Press, 2004.

Hanson, V. D. "Hoplite Battle as Ancient Greek Warfare. When, where, and why?" In H. van Wees, edd., *War and Violence in Ancient Greece*, 201-232. London: Duckworth, 2000.

———. *The Other Greeks. The Family Farm and the Roots of Western Civilization* (2nd ed.). Berkeley and Los Angeles: University of California Press, 1999.

———. *The Western Way of War*. Berkeley and Los Angeles: University of California Press, 1989.

Harris, R. A. *Greek Athletes and Athletics*. Bloomington, IN: Indiana University Press.

———. *Sport in Greece and Rome*. Ithaca: Cornell University Press, 1972.

Herman, G. *Ritualized Friendship and the Greek City*. Cambridge: Cambridge University Press, 1987.

Hodkinson, S. "Lakonian Artistic Production and the Problem of Spartan Austerity." In N. Fisher and H van Wees, edd., *Archaic Greece: New Approaches and New Evidence*, 93-117. London: Duckworth, 1998.

———. "Politics as a determinant of pastoralism: the case of southern Greece." *Rivista di studi liguri* 16 (1992): 139-64.

———. "Imperialist Democracy and Market-Oriented Pastoral Production in Classical Athens." *Anthropolozoologica* 16 (1992): 53-61.

———. "Animal Husbandry in the Greek Polis." In C. R. Whittaker, ed., *Pastoral Economies of Ancient Greece and Rome*. (Cambridge Philological Society, Suppl. Vol. 41), 35-74. Cambridge: Cambridge University Press, 1988.

————. "Social order and the conflict of values in classical Sparta." *Chiron* 13 (1983): 239-81.

Hodkinson, S. and H. Hodkinson. "Mantineia and the Mantinike: settlement and society in a Greek polis." *ABSA* 76 (1981): 239-96.

Hönle, A. *Olympia in der Politik der griechischen Staatenwelt.* Bebenhausen: Lothar Rotsch, 1972.

Howe, T. "Pastoralism, the Delphic Amphiktyony and the First Sacred War: the creation of Apollo's sacred pastures." *Historia* 52 (2003) 129-46.

Hughes, J. D. *Pan's travail: environmental problems of the ancient Greeks and Romans.* Baltimore : Johns Hopkins University Press, 1994.

————. "How the Ancients Viewed Deforestation." *Journal of Field Archaeology* 10 (1983): 437-45.

————. "Deforestation, Erosion, and Forest Management in Ancient Greece and Rome." *Journal of Forest History* 26 (1982): 60-75.

Humphreys, S. *Anthropology and the Greeks.* London: Routledge, 1978.

Isager, S. and J. Skydsgaard. *Ancient Greek Agriculture: An Introduction.* London: Routlege, 1992.

Jameson, M. H. "Sacrifice and Animal Husbandry in Classical Greece." In C. R. Whittaker, ed., *Pastoral Economies of Ancient Greece and Rome* (Cambridge Philological Society, Suppl. Vol. 41), 87-119. Cambridge: Cambridge University Press, 1988.

Jameson, M. H., C. N. Runnels, and T. H. van Andel, edd. *A Greek Countryside. The Southern Argolid from Prehistory to the Present Day.* Stanford: Stanford University Press, 1994.

Jardé, A. *Les Céreales dans l'antiquité greque.* Paris, 1925.

Jones, N. F. *Rural Athens Under the Democracy.* Philadelphia: University of Pennsylvania Press, 2004.

Jost, M. "The Distribution of Sanctuaries in Civic Space in Arkadia." In S. Alcock and R. Osborne, edd., *Placing the Gods. Sanctuaries and Sacred Space in Ancient Greece*, 217-230. Oxford: Oxford University Press, 1994.

Kahrstedt, U. " Delphi und das heilige Land des Apollon." In G. E. Mylonas and D. Raymond, edd., *Studies presented to*

David Moore Robinson II, 749-757. St. Louis: Washington University Press,1953.

Kennedy, R. G. *Rediscovering America*. Boston: Houghton Mifflin, 1990.

Koster, H. A. *The Ecology of Pastoralism in relation to changing patterns of Land Use in the Northwestern Peloponnese*. Diss., University of Pennsylvania,1976.

Kotjabopoulou, E., Y. Hamilakis, P. Halstead, C. Gamble, and V. Elefanti. *Zooarchaeology in Greece. Recent Advances. British School at Athens Studies, 9*. London: The British School at Athens, 2003.

Kurke, L. *The traffic in praise : Pindar and the poetics of social economy*. Ithaca, NY: Cornell University Press, 1991.

———. "The Economy of *Kudos*." In L. Kurke and C. Dougherty, edd. *Cultural Poetics in Archaic Greece*, 131-163. Oxford: Oxford University Press, 1988.

Kyle, D. *Sport and spectacle in the ancient world*. Malden, MA: Blackwell, 2007.

———. *Athletics in Ancient Athens*. Leiden: E. J. Brill, 1987.

Legon, R. P. *Megara*. Ithaca, NY: Cornell University Press, 1981.

Lewis, D. "The Athenian *Rationes Centesimarum*." In M. I. Finley, ed., *Problèmes de la terre en Grèce*, 187-212. Paris: Mouton & Co., 1973.

Lewis, N. "*Leitourgia* and related terms." *GRBS* 3 (1960): 175-184.

Lewthwaite, J. "Plains tales from the hills: transhumance in Mediterranean archaeology." In A. Sheridan and G. Bailey, edd., *Economic Archaeology: Towards an Integration of Ecological and Social Approaches* (BAR S96), 57-66. Oxford: Clarendon Press, 1981.

Lohmann, H. "Antike Hirten in Westkleinasien und der Megaris: zur Archäologie der mediterranen Weidewirtschaft." In Walter Eder and Karl-Joachim Hölkeskamp, edd.,*Volk und Verfassung im vorhellenistichen Griechenland*, 63-88. Stuttgart, Franz Steiner Verlag, 1997.

———. *Atene: Forschungen zu Siedlungs- und Wirtschafts-structur des klassischen Attik, I-II*. Köln: 1993.

————. "Agriculture and Country Life in Classical Attika." In B. Wells, ed., *Agriculture in Ancient Greece*, 29-57. Stockholm: Paul Astrom, 1992.

Luraghi, N. "The Imaginary conquest of the Helots." In N. Luraghi and S. E. Alcock, edd., *Helots and their masters in Laconia and Messenia: histories, ideologies, structures*, 109-141. Cambridge, MA: Harvard University Press, 2003.

MacAldoon, J. J. *Rite, Drama, Festival, Spectacle. Rehearsals Toward a Theory of Cultural Performance*. Philadelphia: University of Pennsylvania Press,1984.

Mantzoulinou-Richards, E. "Demeter Malophoros: The Divine Sheep-Bringer." *AncW* 13 (1986): 15-22.

Marinatos, N. and R. Hågg, edd. *Greek Sanctuaries. New Approaches*. London: Routledge, 1993.

Mauss, M. "Essai sur le don. Forme et raison de l'échange dans les sociétés archaïques." *L'Année Sociologique*, seconde série, 1923-1924 [W. D. Halls, trans., *The gift : the form and reason for exchange in archaic societies*. New York: W.W. Norton, 2000].

McNeil, J. R. *The Mountains of the Mediterranean World: an environmental history*. Cambridge: Cambridge University Press, 1992.

Mee, C. and H. A. Forbes. *A Rough and Rocky Place: The Landscape and Settlement History of the Methana Peninsula, Greece*. Liverpool: Liverpool University Press, 1997.

Meiggs, R. *Trees and Timber in the Ancient World*. Oxford: Clarendon Press, 1982.

Mitchell, L. G. and P. J. Rhodes, edd. *The Development of the polis in Archaic Greece*. London: Routledge, 1997.

Morgan, C. *Early Greek States Beyond the Polis*. London: Routledge, 2003.

————. " The origins of Panhellenism." In N. Marinatos and R. Hägg, edd., *Greek Sanctuaries. New Approaches*, 18-44. London: Routledge, 1993.

————. *Athletes and Oracles*. Cambridge: Cambridge University Press, 1990.

Morris, Ian. "Archaeology and Greek History." In N. Fisher and H. van Wees, edd., *Archaic Greece: New Approaches and New Evidence*, 1-91. London: Duckworth, 1998.

————. "The Art of Citizenship." In S. Langdon, ed., *New Light on a Dark Age*, 9-43. Colombia, MO: University of Missouri Press, 1997.

————. *Death-Ritual and Social Structure in Classical Antiquity.* Cambridge: Cambridge University Press, 1992.

————. "The Use and Abuse of Homer." *Classical Antiquity* 5 (1985): 115-125.

Morris, S. *Daidalos and the Origins of Greek Art.* Princeton: Princeton University Press, 1992.

Munn, M. "The First Excavations at Panakton on the Attic-Boiotian Frontier." *Boeotia Antiqua* 6 (1996): 47-58.

————. "New Light on Panakton and the Attic-Boiotian Frontier." In H. Beister and J. Buckler, edd., *Boiotika. Vortäge vom 5. Internationalen Böotien-Kolloquium zu Ehren von Professor Dr. Siegfried Lauffer,* 231-244. München: Editio Maris, 1989.

Nichols, P. *Aristophanes' Novel Forms: The Political Role of Drama.* London: Minerva,1998.

Nixon, L. and S. Price. "The Diachronic Analysis of Pastoralism through Comparative Variables." *ABSA* 96 (2001): 395-424.

Ober, J. *Mass and Elite in Democratic Athens.* Princeton: Princeton University Press, 1989.

Osborne, R. "'Is it a Farm?' The Definition of Agricultural Sites and Settlements in Ancient Greece." In B. Wells, ed., *Agriculture in Ancient Greece,* 21-28. Stockholm: Paul Astrom, 1992.

————. *The Classical Landscape with Figures.* London: George Philip, 1987.

————. *Demos: The Discovery of Classical Attika.* Cambridge: Cambridge University Press, 1985.

Parker, R. *Athenian Religion: A History.* Oxford: Clarendon Press, 1996.

Parker, V. *Untersuchen zum Lalantischen Krieg und verwandten Problemen der frühgriechischen Geschichte.* Stuttgart: Franz Steiner Verlag, 1997.

Payne, S. "Zoo-archaeology in Greece: A Reader's Guide." In Nancy C. Wilkie and William D. E. Coulson, edd., *Contributions to Aegean Archaeology. Studies in Honor of William A. McDonald,* 211-44. Minneapolis: University of Minnesota Press, 1985.

Poliakoff, M. B. *Combat Sports in the Ancient World.* New Haven: Yale University Press, 1987.

Polyani, K. *The great transformation.* New York: Farrar & Rinehart, 1944.

Pomeroy, S. B., S. M. Burstein, W. Donlan, and J. Roberts. *Ancient Greece: a political, social, and cultural history.* Oxford: Oxford University Press, 1999.

Ponting, C. *A Green History of the World: the environment and the collapse of great civilizations.* London: Sinclair-Stevenson, 1991.

Qviller, B. "The Dynamics of the Homeric Society." *Symbolae Osloenses* 56 (1981): 109-55.

Raaflaub, K. "A Historian's Headache. How to read 'Homeric Society'?" In N. Fisher and H. van Wees, edd., *Archaic Greece: New Approaches and New Evidence*, 169-193. London: Duckworth, 1998.

————. "Soldiers, Citizens and the Evolution of the Early Greek *Polis*." In L. G. Mitchell and P. J. Rhodes, edd., *The Development of the polis in Archaic Greece*, 49-59. London: Routledge, 1997.

Rackham, O. *Trees, wood and timber in Greek history.* Oxford: Leopard's Press, 2001.

————. "Ecology and Pseudo-Ecology." In G. Shipley and J. Salmon, eds., *Human Landscapes in Classical Antiquity*, 16-43. London and New York: Routledge, 1996.

————. "Ancient landscapes." In O. Murray and S. Price, eds., *The Greek City: From Homer to Alexander*, 85-111. Oxford: Clarendon Press, 1990.

————. "The countryside: history and pseudo-history." *The Historian* 14 (1987): 13-17.

————. "Observations on the Historical Ecology of Boeotia." *ABSA* 78 (1983): 291-351.

————. "Land use and the Native Vegetation of Greece." In M. Bell and S. Limbrey, edd., *Archaeological Aspects of a Woodland Ecology*, 177-98. Oxford: BAR International Series, 1982.

Rackham, O. and J. Moody. *The Making of the Cretan Landscape.* Manchester: Manchester University Press, 1996.

Reese, D. "Recent Work in Greek Zooarchaeology." In P. Nick Kardulias, ed., *Beyond the Site. Regional Studies in the*

Aegean Area, 191-223. Lanham, MD: University Press of America, 1994.

Richter, W. *Die Landwirtschaft im homerischen Zeitalter. Mit einem Beitrag: Landwirtschaftlische Geräte, von Wolfgang Schiering* (Archaeologia Homerica II, H). Göttingen, 1968.

Robert, L. *Études épigraphiques et philologiques*. Paris, 1938.

————. "Epitaphe d'un berger á Thasos." *Hellenika* 7 (1949): 152-160.

————. "Les chèvres d'Herakleia." *Hellenika* 7 (1949): 161-170.

————. "Monnaies d'Olympos." *Hellenika* 10 (1955): 185-186.

Rosen, R. M. and I. Sluiter, edd. *City, Countryside, and the Spatial Organization of Value in Classical Antiquity. Mnemosyne Supplements* 279. Leiden: Brill, 2006.

Rosivach, V. *The System of Public Sacrifice in Fourth-Century Athens.* American Classical Studies 34. Atlanta: Scholars Press, 1994.

Roux, G. *L'Amphictionie, Delphes et le temple d'Apollo au VIe siècle.* Lyons: Maison de l'Orient, 1979.

Sallares, R. *The Ecology of the Ancient Greek World.* Ithaca, New York: Cornell University Press, 1991.

Sartre, M. "Aspects économiques et aspects religieux de la frontière dans les cités grecques." *Ktema* 4 (1979): 213-24.

Scheidel, W. and S. von Reden, edd. *The Ancient Economy.* London: Rouledge, 2002.

Sealey, R. *Demosthenes and his time: a study in defeat.* Oxford: Oxford University Press, 1993.

————. *A History of the Greek City States 700-338 B.C.* Berkeley and Los Angeles: University of California Press, 1976.

Semple, E. C. "The Influence of Geographic Conditions upon Ancient Mediterranean Stock-Raising." *Annals of the Association of American Geographers* 12 (1922): 3-38.

————. *The geography of the Mediterranean region and its relation to ancient history.* New York: H. Holt and Company, 1932.

Sheehan, M. C. *The postglacial vegetational history of the Argolid Peninsula, Greece.* Diss., University of Indiana, 1979.

Skydsgaard, J. "Transhumance in Ancient Greece." In C. R. Whittaker, ed., *Pastoral Economies of Ancient Greece and Rome* (Cambridge Philological Society, Suppl. Vol. 41), 75-86. Cambridge: Cambridge University Press, 1988.

Snodgrass, A. M. *An Archaeology of Greece: The Present and Future Scope of a Discipline.* Berkeley and Los Angeles: University of California Press, 1989.

————. *Archaic Greece: The Age of Experiment.* Berkeley and Los Angeles: University of California Press, 1980.

————. *The Dark Age of Greece.* Edinburgh: Edinburgh University Press, 1971.

————. "The Hoplite Reform and History." *JHS* 85(1965): 110-22.

Sonnini, C. S. *Voyage en Grèce et en Turquie fait par ordre de Louis XVI.* Paris, 1801.

Sourvinou-Inwood, C. "Early sanctuaries, the eight century and ritual space. Fragments of a discourse." In N. Marinatos and R. Hagg, edd., *Greek Sanctuaries*, 1-17. London: Routledge, 1993.

————. *"Reading" Greek culture: texts and images, rituals and myths.* Oxford: Oxford University Press, 1991.

Spence, I. G. *The Cavalry of Classical Greece.* Oxford: Oxford University Press, 1993.

Sprague, R. K. *The Older Sophists.* Columbia, SC: University of Missouri Press, 1972.

Stanton, G. "Some Attic Inscriptions." *ABSA* 92 (1997): 178-204.

————. "Some Attic Inscriptions." *ABSA* 79 (1984): 289-306.

Thomas, C. and C. Conant. *Citadel to City State. The transformation of Greece, 1200-700 B.C.E.* Bloomington, ID: Indiana University Press, 1999.

Tomlinson, R. A. *Argos and the Argolid.* Ithaca, NY: Cornell University Press, 1972.

van Andel, T. H. and C. Runnels. *Beyond the Acropolis. A Rural Greek Past.* Stanford: Stanford University Press, 1987.

van de Maele, S. "L' Orgas Eleusinienne: Etude Topographique." *Melanges Edouard Delabeque*, 417-433. Marseilles,1983.

van Wees, H. *Greek Warfare. Myths and Realities.* London: Duckworth, 2004.

————. "Megara's Mafiosi: Timocracy and Violence in Theognis." In R. Brock and S. Hodkinson, *Alternatives to Athens. Varieties of Political Organization and Community in Ancient Greece*, 52-67. Oxford: Oxford University Press, 2001.

————. "The Development of the Hoplite Phalanx. Iconography and reality in the seventh century." In H. van Wees, ed.,*War and Violence in Ancient Greece*, 125-166. London: Duckworth, 2000.

————. "Greeks Bearing Arms. The state, the leisure class, and the display of weapons in archaic Greece." In N. Fisher and H. van Wees, edd., *Archaic Greece: New Approaches and New Evidence*, 333-377. London: Duckworth, 1998.

————. *Status Warriors: War, Violence and Society in Homer and History*. Amsterdam: J.C. Gieben, 1992.

Veyne, P. *Bread and circuses: historical sociology and political pluralism*. London: Penguin, 1990.

Vidal-Naquet, P. "La tradition de l'hoplite athénien." In J.-P. Vernant, ed., *Problèmes de la guerre en Grèce ancienne*, 161-181. Paris: La Haye, 1968.

Voyatzis, M. *The Early Sanctuary of Athena Alea at Tegea. And Other Archaic Sanctuaries in Arcadia*. Göteborg: Paul Astrom, 1990.

Westlake, H. D. *Thessaly in the Fourth Century B.C.* Oxford: Oxford University Press, 1935.

White, K. D. *Roman Farming*. Ithaca, NY: Cornell University Press, 1970.

————. "Wheat Farming in Roman Times." *Antiquity* 37 (1963): 209.

Whitehead, D. "Competitive Outlay and Community Profit: Filotimia in Classical Athens." *Classica et Mediaevalia* 34 (1983): 55-74.

Will, Éduard. "Aux origines du régime foncier grec. Homère, Hésiode et l'arrière-plan Mycénien." *Révue des Études Anciennes* 59 (1957): 5-50.

Winkelstern, K. *Die Schweinezucht im klassischen Altertum*. Diss., Giessen, 1933.

Wright, H. E. "Vegetation History." In W. A. McDonald and G. Rapp, edd. *The Minnesota Messenia Expedition*, 188-99. Minneapolis: University of Minnesota Press, 1972.

Worthington, I. ed. *Demosthenes: statesman and orator.* London: Routledge, 2000.

Young, D. C. *The Olympic Myth of Greek Amateur Athletics.* Chicago: Ares Press, 1984.

Zeissig, K. *Die Rinderzucht im alten Griechenland.* Diss., Giessen, 1934.